MCQs Series for Life Sciences

(Volume 2)

(Cell & Tissue Culture and Microbiology)

Authored by

Maddaly Ravi

*Department of Human Genetics, Sri Ramachandra Medical College & Research
Institute, Porur, Chennai - 600 116, India*

General:

1. Any dispute or claim arising out of or in connection with this License Agreement or the Work (including non-contractual disputes or claims) will be governed by and construed in accordance with the laws of the U.A.E. as applied in the Emirate of Dubai. Each party agrees that the courts of the Emirate of Dubai shall have exclusive jurisdiction to settle any dispute or claim arising out of or in connection with this License Agreement or the Work (including non-contractual disputes or claims).
2. Your rights under this License Agreement will automatically terminate without notice and without the need for a court order if at any point you breach any terms of this License Agreement. In no event will any delay or failure by Bentham Science Publishers in enforcing your compliance with this License Agreement constitute a waiver of any of its rights.
3. You acknowledge that you have read this License Agreement, and agree to be bound by its terms and conditions. To the extent that any other terms and conditions presented on any website of Bentham Science Publishers conflict with, or are inconsistent with, the terms and conditions set out in this License Agreement, you acknowledge that the terms and conditions set out in this License Agreement shall prevail.

Bentham Science Publishers Ltd.
Executive Suite Y - 2
PO Box 7917, Saif Zone
Sharjah, U.A.E.
Email: subscriptions@benthamscience.org

BENTHAM SCIENCE

CONTENTS

PREFACE

The MCQs series for Life Sciences, Volume I published in 2014, covered five areas of Life Sciences, Immunology, Biochemistry, Cell Biology Developmental Biology, Genetics and Molecular Biology. The series is now complete with the Volume II which covers several topics in the areas of Cell Cultures and Microbiology. Both the volumes aim to address the needs of students in all the major areas directly related to Life Sciences for not only preparing for competitive examinations, but also as an augment for understanding the subjects better. The subjects have been comprehensively covered, while avoiding redundancy and 380 and 547 MCQs for Cell Cultures and Microbiology respectively are presented in this Volume II. As with the Volume II, all MCQs are presented in a simple language with straight forward multiple choices and answers.

I am also happy that one chapter in this Volume "Chapter 14. Microbial Diseases" in Part 2 – Microbiology has been contributed by one of our students Ms. Shwetha Krishna who is currently pursuing her M.Sc. in Human Genetics at our College.

I sincerely wish that this Volume II of the MCQs series in Life Sciences will be useful for students to improve their performance in the regular course work examinations/tests as well as for those who are preparing for competitive/qualifying/entrance examinations/tests. I once again look forward to your feedback which will be useful for me to address shortcomings, if any in these two Volumes.

CONSENT FOR PUBLICATION

Not applicable.

CONFLICT OF INTEREST

The author confirms that this eBook contents have no conflict of interest.

ACKNOWLEDGEMENTS

I sincerely acknowledge all my teachers, students and my family for their support and for what I have learnt from them. My sincere acknowledgements also to the team (Mr. Mahmood Alam, Mr. Shehzad Naqvi and Ms. Fariya Zulfiqar) of Bentham eBooks for the utmost professional manner in which both Volume I and II have been handled at each of the stages of editorial, review and publishing process. Also, the inputs and comments from the expert reviewers were extremely useful to render the manuscripts as flaw-less as possible, and I sincerely thank them for the valuable inputs.

Maddaly Ravi
Department of Human Genetics
Sri Ramachandra Medical College & Research Institute
Porur, Chennai - 600 116
India
Tel: +91 98 41486363
Fax: +91 44 24767008
E-mail: mravi@sriramachandra.edu.in

MCQs Series for Life Sciences

(Volume 2)

(Cell & Tissue Culture and Microbiology)

CHAPTER 1

Types of Animal Cell Cultures and Major Discoveries

1. Primary cell cultures are obtained from
   a. Tissues or organs of life forms
   b. From continuous cell lines
   c. From cancer cell lines
   d. From transformed cells

2. Cells for primary cultures are obtained from their source tissue by
   a. Enzymatic or mechanical dissociation
   b. By chemical digestion
   c. By increased temperatures
   d. By transformation

3. Adherent cells are also known as
   a. Suspension cells
   b. Aggregates
   c. Anchorage dependent cells
   d. Transformed cells

4. Adherent cells typically form
   a. Single cell cultures
   b. Suspension cells
   c. Aggregates
   d. Monolayers

5. Non-adherent cells typically form
   a. Single cell cultures
   b. Suspension cultures
   c. Aggregates
   d. Monolayers

6. Secondary cell cultures are obtained from
   a. Stem cells
   b. Source tissues or organs
   c. Primary cultures
   d. 3D aggregates

7. Subculturing of cells in cultures is also known as
   a. Typsinization
   b. Passaging
   c. Acclimatization
   d. Conditioning

8. Cell lines can be
   a. Finite
   b. Continuous
   c. Finite or continuous
   d. Transforming cells

9. Contact inhibition property is typical of
   a. Finite cell lines
   b. Continuous cell lines
   c. Suspension cell lines
   d. 3D aggregates

10. Explant cultures are a type of
    a. Primary cells
    b. Secondary cells
    c. Transformed cells
    d. Continuous cell lines

11. The type of cultures that typically dominated by fibroblasts are
    a. 3D cultures
    b. Co-cultures
    c. Explant cultures
    d. Transformed cell cultures

12. Continuous cell lines are obtained by
    a. Fusion
    b. Transformation
    c. Passaging
    d. Selection

13. 2-dimensional cell cultures include
    a. Adherent cultures
    b. Suspension cells
    c. Both adherent and suspension cultures
    d. Aggregates

14. Adherent cell lines usually form
    a. Monolayers
    b. Stratified layers
    c. Necrotic cores
    d. Aggregates

15. The cell types that retain highest biotransformation ability are
    a. Continuous cell lines
    b. Finite cell lines
    c. Primary cells
    d. Secondary cell lines

16. The cell types that retain highest tissue-distinct functions are
    a. Continuous cell lines
    b. Finite cell lines
    c. Primary cells
    d. Secondary cell lines

17. The cells of the circulatory system are cultured as
    a. Monolayers
    b. Adherent cells
    c. Suspension cells
    d. Aggregates

18. The type of cultured cells with elongated bipolar morphology are typically
    a. Epithelial-like
    b. Fibroblastic
    c. Lymphoblast-like
    d. Spheroid-like

19. The type of cultured cells with polygonal or other regular morphology are typically
    a. Epithelial-like
    b. Fibroblastic
    c. Lymphoblast-like

d. Spheroid-like

20. The type of cultured cells with spherical morphology are typically
    a. Epithelial-like
    b. Fibroblastic
    c. Lymphoblast-like
    d. Spheroid-like

21. Cell aggregates are typically formed by
    a. Monolayer cell cultures
    b. Suspension cell cultures
    c. 3D cell cultures
    d. Primary cell cultures

22. A distinct tissue-like morphology can be obtained by culturing cells as
    a. Monolayer cell cultures
    b. Suspension cell cultures
    c. 2D cell cultures
    d. 3D cell cultures

23. The type of cell cultures that required a matrix or scaffold is
    a. 2D cultures
    b. 3D cultures
    c. Explants
    d. Transformed cells

24. Typically, 3D cell cultures are obtained by
    a. Preventing cells from attaching to the culture vessels
    b. Expanding explant cultures
    c. Transforming cancerous cells
    d. Limited passaging of primary cells

25. The type of cell cultures that are considered closer to *in vivo* conditions are
    a. Passaged cell lines
    b. 2D cultures
    c. 3D cultures
    d. Suspension cell cultures

26. Organoids are typically obtained by using
    a. Stem cells
    b. Cancer cells

  c. Transformed cells

  d. Tissues

27. The possibility that tissues can be maintained physiologically active after the death of an organism was first proposed by

  a. Jolly

  b. Enders

  c. Robbins

  d. Alec Issacs

28. The first *in vitro* cell growth was demonstrated by culturing frog embryo nerve fiber by

  a. Carrel in 1923

  b. Ebling in 1912

  c. Harrison in 1907

  d. Jones in 1916

29. Chick embryo explants were used for the first time to demonstrate contraction of heart muscle by

  a. Carrel and Burrows

  b. Ebling and Carrel

  c. Borrows and Harrison

  d. Jones and Rous

30. Utilization of trypsin for explant sub culturing was first demonstrated by

  a. Carrel and Burrows

  b. Ebling and Carrel

  c. Borrows and Harrison

  d. Jones and Rous

31. Fibroblast cell lines were first subcultured by

  a. Carrel and Burrows

  b. Ebling and Carrel

  c. Borrows and Harrison

  d. Jones and Rous

32. The introduction of antibiotics in cell cultures to prevent contamination was first demonstrated by

  a. Eagle in 1955

  b. Gey in 1952

  c. Dulbecco in 1952

d. Keilova in 1948

33. The utilization of trypsin for passaging cell cultures was first demonstrated by
    a. Eagle in 1955
    b. Gey in 1952
    c. Dulbecco in 1952
    d. Keilova in 1948

34. The first human cancer cell line HeLa was first established in the year(s)
    a. 1952-1955
    b. 1948-1950
    c. 1950-1952
    d. 1946-1948

35. The first human cancer cell line HeLa was first established by
    a. Eagle in 1955
    b. Gey in 1952
    c. Dulbecco in 1952
    d. Keilova in 1948

36. Defined media for cell cultures were introduced by
    a. Eagle in 1955
    b. Gey in 1952
    c. Dulbecco in 1952
    d. Keilova in 1948

37. The cells in culture which have a limited life span were first demonstrated by
    a. Moore in 1967
    b. Sorieul and Ephrussi in 1961
    c. Hayflick and Moorhead in 1961
    d. Borrows and Harrison in 1955

38. Fusion of two cells by somatic cell hybridization was first demonstrated by
    a. Moore in 1967
    b. Sorieul and Ephrussi in 1961
    c. Mayflick and Moorhead in 1961
    d. Borrows and Harrison in 1955

39. The lymphoblastoid cell lines were first established by
    a. Moore in 1967
    b. Gey in 1952

   c. Moorhead in 1961
   d. Ephrussi in 1961

40. The hanging-drop method was first demonstrated by
   a. Ross Harrison
   b. Leo Loeb
   c. Alexis Carrel
   d. Montrose Burrows

41. Culture of blood cells in serum and plasma was first demonstrated by
   a. Lewis in 1911
   b. Jolly in 1903
   c. Loeb in 1897
   d. Roux in 1885

42. Cell divisions in cultured leucocytes was first observed by
   a. Lewis in 1911
   b. Jolly in 1903
   c. Carrel in 1923
   d. Roux in 1885

43. A detailed study of mitosis using cultured chick embryo cells was first performed by
   a. Rivera in 1927
   b. Carrel in 1923
   c. Rous in 1916
   d. Burrows in 1910

44. The first liquid medium for culturing cells *in vitro* was first formulated by
   a. Lewis and Lewis in 1911
   b. Carrel and Baker in 1923
   c. Hayflick and Moorhead in 1961
   d. Harris and Watkins in 1965

45. The importance and demonstration of antiseptic techniques in cell cultures was first demonstrated by
   a. Lewis in 1911
   b. Carrel in 1913
   c. Hayflick in 1961
   d. Watkins in 1965

46. T-flasks for culturing cells were first introduced by
    a. Lewis and Lewis in 1911
    b. Carrel and Baker in 1923
    c. Hayflick and Moorhead in 1961
    d. Harris and Watkins in 1965

47. Cell cultures were used for the first time to develop (Vaccinia) vaccines by
    a. Carrel and Baker in 1923
    b. Harris and Watkins in 1965
    c. Earle and Enders in 1948
    d. Carrel and Rivera in 1927

48. The roller tube method for cell cultures was introduced by
    a. Gey in 1933
    b. Jolly in 1903
    c. Roux in 1885
    d. Watkins in 1965

49. The first chemically defined culture medium was
    a. DMEM
    b. RPMI
    c. CMRL 1066
    d. Hybrimax

50. The utilization of confluent monolayers for developing plaque assay to study animal viruses was first demonstrated by
    a. Gey in 1952
    b. Dulbecco in 1952
    c. Hayflick in 1961
    d. Littlefield in 1964

51. Contact inhibition in cultured cells was first explained by
    a. Littlefield in 1964
    b. Roux in 1885
    c. Carrel in 1923
    d. Abercrombie in 1954

52. The HAT medium was introduced as a selection medium by
    a. Littlefield in 1964
    b. Roux in 1885
    c. Carrel in 1923

    d. Abercrombie in 1954

53. Serum-free medium for cell cultures was first introduced by
    a. Roux in 1885
    b. Dulbecco in 1952
    c. Eagle in 1955
    d. Ham in 1965

54. Hybridomas were first generated by
    a. Kohler and Milstein in 1975
    b. Carrel and Baker in 1923
    c. Earle and Enders in 1948
    d. Borrows and Harrison in 1955

CHAPTER 2

Animal Cell Culture Media and Supplements

1. The media used for cell cultures during early cell culture technique development did not include
    a. Lymph
    b. Serum
    c. Embryo extracts
    d. Saline

2. The development of chemically defined medium was augmented by the understanding of
    a. Media components
    b. Nutritional biochemistry
    c. Types of media available
    d. Types of cells to grow

3. The first chemically defined medium that was first developed in 1955 was
    a. Eagle's Basal Medium
    b. Eagle's Minimum Essential Medium
    c. Dulbecco's Modified Eagle's Medium
    d. Ham's F12 Medium

4. The pH indicator most commonly used in cell culture medium is
    a. Trypan blue
    b. Coomassie blue
    c. Phenol red
    d. Propidium iodide

5. The colour of phenol red at the pH 7.4 is
    a. Pink
    b. Yellow
    c. White
    d. Red

6. The colour of phenol red at the pH 6.5 is
   a. Pink
   b. Yellow
   c. White
   d. Red

7. The colour of phenol red at the pH 7.6 is
   a. Pink
   b. Yellow
   c. White
   d. Red

8. The production of endogenous CO_2 by cells being cultured can be enabled by the supplementation of medium with
   a. Phenol red
   b. Serum
   c. Pyruvate
   d. Bicarbonate

9. The buffering action in cell culture medium is facilitated by the addition of
   a. Sodium bicarbonate
   b. Fetal bovine serum
   c. Albumin
   d. Phenol red

10. The requirement of CO_2 for cells cultured in open vessels is
    a. High
    b. Low
    c. Nil
    d. Not an important consideration

11. The requirement of CO_2 for cells cultured in high concentration is
    a. High
    b. Low
    c. Nill
    d. Not an important consideration

12. The requirement of CO_2 for cells cultured in low concentration is
    a. High
    b. Low
    c. Nil

d. Not an important consideration

13. The requirement of CO_2 for cells cultured in sealed vessels is
    a. High
    b. Low
    c. Nil
    d. Not an important consideration

14. A strong buffering of liquid cell culture medium can be obtained by the supplementation of
    a. Bicarbonate
    b. HEPES
    c. CO_2
    d. NaOH

15. The salts provided in the cell culture liquid medium mainly function for
    a. High nutrition
    b. pH
    c. Osmolality
    d. Viscosity

16. The function of serum for cell culture does not include
    a. Promotion of cell proliferation
    b. Antitrypsin activity
    c. To increase osmolarity
    d. For providing adhesion factors

17. The type of serum with low levels of polyamine oxidase is
    a. Calf serum
    b. Horse serum
    c. Human serum
    d. Fetal bovine serum

18. The type of serum that should be supplemented as the most appropriate for culturing 3T3 cells is
    a. Calf serum
    b. Fetal bovine serum
    c. Horse serum
    d. Human serum

19. The type of serum that should be supplemented as the most appropriate for culturing chick embryo fibroblasts is
    a. Calf serum

b. Fetal bovine serum
c. Horse serum
d. Human serum

20. The type of serum that should be supplemented as the most appropriate for culturing Chinese hamster ovary (CHO) cells is
a. Calf serum
b. Fetal bovine serum
c. Horse serum
d. Human serum

21. The type of serum that should be supplemented as the most appropriate for culturing chondrocytes is
a. Calf serum
b. Fetal bovine serum
c. Horse serum
d. Human serum

22. The type of serum that should be supplemented as the most appropriate for culturing most continuous cell lines is
a. Calf serum
b. Fetal bovine serum
c. Horse serum
d. Human serum

23. The type of serum that should be supplemented as the most appropriate for culturing endothelial cells is
a. Calf serum
b. Fetal bovine serum
c. Horse serum
d. Human serum

24. The type of serum that should be supplemented as the most appropriate for culturing fibroblasts is
a. Horse serum
b. Fetal bovine serum
c. Calf serum
d. Human serum

25. The type of serum that should be supplemented as the most appropriate for culturing glial cells is

a. Horse serum
b. Fetal bovine serum
c. Calf serum
d. Human serum

26. The type of serum that should be supplemented as the most appropriate for culturing glioma cells is
   a. Horse serum
   b. Fetal bovine serum
   c. Calf serum
   d. Human serum

27. The type of serum that should be supplemented as the most appropriate for culturing HeLa cells is
   a. Horse serum
   b. Fetal bovine serum
   c. Calf serum
   d. Human serum

28. The type of serum that should be supplemented as the most appropriate for culturing hematopoietic cells is
   a. Horse serum
   b. Fetal bovine serum
   c. Calf serum
   d. Human serum

29. The type of serum that should be supplemented as the most appropriate for culturing human diploid fibroblasts is
   a. Horse serum
   b. Fetal bovine serum
   c. Calf serum
   d. Human serum

30. The type of serum that should be supplemented as the most appropriate for culturing human leukemia cells is
   a. Horse serum
   b. Fetal bovine serum
   c. Calf serum
   d. Human serum

31. The type of serum that should be supplemented as the most appropriate for

culturing human tumor cells is
a. Horse serum
b. Calf serum
c. Fetal bovine serum
d. Human serum

32. The type of serum that should be supplemented as the most appropriate for culturing keratinocytes is
a. Horse serum
b. Fetal bovine serum
c. Calf serum
d. Human serum

33. The type of serum that should be supplemented as the most appropriate for culturing human lymphoblastoid cells is
a. Horse serum
b. Fetal bovine serum
c. Calf serum
d. Human serum

34. The type of serum that should be supplemented as the most appropriate for culturing mammary epithelial cells is
a. Horse serum
b. Fetal bovine serum
c. Calf serum
d. Human serum

35. The type of serum that should be supplemented as the most appropriate for culturing melanocytes is
a. Horse serum
b. Fetal bovine serum
c. Calf serum
d. Human serum

36. The type of serum that should be supplemented as the most appropriate for culturing melanoma is
a. Fetal bovine serum
b. Horse serum
c. Calf serum
d. Human serum

37. The type of serum that should be supplemented as the most appropriate for culturing mouse embryo fibroblasts is
    a. Fetal bovine serum
    b. Horse serum
    c. Calf serum
    d. Human serum

38. The type of serum that should be supplemented as the most appropriate for culturing mouse myeloma cells is
    a. Fetal bovine serum
    b. Horse serum
    c. Calf serum
    d. Human serum

39. The type of medium that is most appropriate for culturing 3T3 cells is
    a. DMEM
    b. Eagle's MEM
    c. RPMI 1640
    d. Ham's F12

40. The type of medium that is most appropriate for culturing chick embryo fibroblasts is
    a. DMEM
    b. Eagle's MEM
    c. RPMI 1640
    d. Ham's F12

41. The type of medium that is most appropriate for culturing CHO cells is
    a. DMEM
    b. Eagle's MEM
    c. RPMI 1640
    d. L15

42. The type of medium that is most appropriate for culturing chondrocytes is
    a. DMEM
    b. Eagle's MEM
    c. RPMI 1640
    d. Ham's F12

43. The type of medium that is most appropriate for culturing most continuous cell lines is

a. DMEM
b. L15
c. RPMI 1640
d. Ham's F12

44. The type of medium that is most appropriate for culturing endothelial cells is
    a. DMEM
    b. Eagle's MEM
    c. RPMI 1640
    d. Ham's F12

45. The type of medium that is most appropriate for culturing fibroblasts is
    a. DMEM
    b. Eagle's MEM
    c. RPMI 1640
    d. Ham's F12

46. The type of medium that is most appropriate for culturing glial cells is
    a. MEM, DMEM/F12
    b. Eagle's MEM
    c. RPMI 1640
    d. Ham's F12

47. The type of medium that is most appropriate for gliomas is
    a. MEM, DMEM/F12
    b. Eagle's MEM
    c. RPMI 1640
    d. Ham's F12

48. The type of medium that is most appropriate for culturing HeLa is
    a. DMEM
    b. Eagle's MEM
    c. RPMI 1640
    d. Ham's F12

49. The type of medium that is most appropriate for culturing hematopoietic cells is
    a. DMEM
    b. Eagle's MEM
    c. RPMI 1640
    d. Ham's F12

50. The type of medium that is most appropriate for culturing human diploid fibroblasts is
    a. DMEM
    b. Eagle's MEM
    c. RPMI 1640
    d. Ham's F12

51. The type of medium that is most appropriate for culturing human leukemia cells is
    a. DMEM
    b. Eagle's MEM
    c. RPMI 1640
    d. Ham's F12

52. The type of medium that is most appropriate for culturing keratinocytes is
    a. DMEM
    b. Eagle's MEM
    c. RPMI 1640
    d. αMEM

53. The type of medium that is most appropriate for culturing human lymphoblastoid cells is
    a. DMEM
    b. Eagle's MEM
    c. RPMI 1640
    d. Ham's F12

54. The type of medium that is most appropriate for culturing mammary epithelial cells is
    a. DMEM, MEM
    b. Eagle's MEM/F12
    c. RPMI 1640, DMEM/F12
    d. Ham's F12

55. The type of medium that is most appropriate for culturing melanocytes is
    a. F12
    b. M199
    c. RPMI 1640
    d. Ham's F12

56. The type of medium that is most appropriate for culturing melanoma cells is

    a. DMEM/F12, MEM
    b. Eagle's MEM, Ham's F12
    c. RPMI 1640, GMEM
    d. Ham's F12, M199

57. The type of medium that is most appropriate for culturing mouse embryo fibroblasts is
    a. DMEM/F12, MEM
    b. Eagle's MEM
    c. RPMI 1640, GMEM
    d. Ham's F12

58. The type of medium that is most appropriate for culturing mouse leukemia cells is
    a. DMEM/F12, MEM
    b. Eagle's MEM, Ham's F12
    c. RPMI 1640, Fisher's
    d. Ham's F12, M199

59. The type of medium that is most appropriate for culturing mouse erythroleukemia cells is
    a. DMEM/F12, RPMI 1640
    b. Eagle's MEM, Ham's F12
    c. RPMI 1640, Fisher's
    d. Ham's F12, M199

60. The type of medium that is most appropriate for culturing mouse myeloma cells is
    a. DMEM, RPMI 1640
    b. Eagle's MEM, Ham's F12
    c. RPMI 1640, Fisher's
    d. Ham's F12, M199

61. The type of medium that is most appropriate for culturing mouse neuroblastoma cells is
    a. Ham's F12, M199
    b. Eagle's MEM, Ham's F12
    c. RPMI 1640, Fisher's
    d. DMEM, DMEM/F12

62. The type of medium that is most appropriate for culturing neurons cells is

a. Ham's F12
b. Eagle's MEM
c. RPMI 1640
d. DMEM

63. The type of medium that is most appropriate for culturing skeletal muscle cells is
a. Ham's F12, M199
b. Eagle's MEM, Ham's F12
c. RPMI 1640, Fisher's
d. DMEM, F12

64. Inactivation of serum that is to be supplemented to liquid culture media is achieved by
a. Heating
b. Filtration
c. UV exposure
d. Dilution

65. The serum to be supplemented to liquid culture medium is inactivated by exposing to
a. 37°C
b. 56°C
c. 76°C
d. 93°C

66. Spent medium from cell cultures that is supplemented to boost cultures of difficult cells is known as
a. Complete medium
b. Serum-free medium
c. Incomplete medium
d. Conditioned medium

67. The disadvantages of using serum as a supplement for liquid cell culture medium do not include
a. Quality control for batch-to-batch variation
b. Shelf life
c. Presence of growth factors
d. Physiological variability

68. The serum-free medium most ideally suited to culture human lung fibroblasts

is
a. MCDB 110
b. MCDB 131
c. MCDB 302
d. MCDB 404

69. The serum-free medium most ideally suited to culture human vascular endothelium is
a. MCDB 110
b. MCDB 153
c. MCDB 131
d. MCDB 404

70. The serum-free medium most ideally suited to culture human mammary epithelium is
a. MCDB 153
b. MCDB 170
c. MCDB 131
d. MCDB 302

71. The serum-free medium most ideally suited to culture chick embryo fibroblasts is
a. MCDB 202
b. MCDB 302
c. MCDB 402
d. MCDB 404

72. The serum-free medium most ideally suited to culture CHO cells is
a. MCDB 202
b. MCDB 302
c. MCDB 402
d. MCDB 404

73. The serum-free medium most ideally suited to culture 3T3 cells is
a. MCDB 202
b. MCDB 302
c. MCDB 402
d. MCDB 404

74. The serum-free medium most ideally suited to culture prostatic epithelial is
a. MCDB 202

   b. WAJC 404
   c. MCDB 402
   d. MCDB 302

75. The serum-free medium most ideally suited to culture keratinocytes is
   a. MCDB 131
   b. MCDB 170
   c. MCDB 153
   d. MCDB 404

76. The serum-free medium most ideally suited to culture lymphoid cells is
   a. Iscove's medium
   b. DMEM
   c. RPMI
   d. F12

77. The serum-free medium most ideally suited to culture bronchial epithelial cells is
   a. MCDB 131
   b. MCDB 170
   c. Iscove's medium
   d. LHC-9

78. The basal medium most suited to prepare supplemented basal medium for culturing human small-cell lung carcinoma cells is
   a. RPMI 1640
   b. DMEM
   c. DMEM/F12
   d. MCDB153

79. The basal medium most suited to prepare supplemented basal medium for culturing human non-small-cell lung carcinoma cells is
   a. RPMI 1640
   b. DMEM
   c. DMEM/F12
   d. MCDB153

80. The basal medium most suited to prepare supplemented basal medium for culturing human lung adenocarcinoma cells is
   a. RPMI 1640
   b. DMEM

    c. DMEM/F12
    d. MCDB153

81. The basal medium most suited to prepare supplemented basal medium for culturing rat glial cells is
    a. RPMI 1640
    b. DMEM
    c. DMEM/F12
    d. MCDB153

82. The basal medium most suited to prepare supplemented basal medium for culturing dog kidney MDCK cells is
    a. RPMI 1640
    b. DMEM
    c. DMEM/F12
    d. MCDB153

83. The basal medium most suited to prepare supplemented basal medium for culturing pig kidney LLC-PK cells is
    a. RPMI 1640
    b. DMEM
    c. DMEM/F12
    d. MCDB153

84. The basal medium most suited to prepare supplemented basal medium for culturing human parotid cells is
    a. RPMI 1640
    b. DMEM
    c. DMEM/F12
    d. MCDB153

85. The basal medium most suited to prepare supplemented basal medium for culturing human prostate cells is
    a. RPMI 1640
    b. αMEM/F12
    c. DMEM/F12
    d. MCDB153

86. The basal medium most suited to prepare supplemented basal medium for culturing human melanocyte cells is
    a. M199

    b. αMEM/F12

    c. DMEM/F12

    d. MCDB153

87. The mitogen that is supplemented to the medium as specific only for T lymphocytes is
    a. Phytohaemagglutinin (PHA)
    b. Pokeweed mitogen (PWM)
    c. Lipopolysaccharide (LPS)
    d. Protein A

88. The mitogen that is supplemented to the medium as specific only for T lymphocytes is
    a. Protein A
    b. Pokeweed mitogen (PWM)
    c. Lipopolysaccharide (LPS)
    d. Concanavalin A

89. The mitogen that is supplemented to the medium as specific only for B lymphocytes is
    a. Phytohaemagglutinin (PHA)
    b. Pokeweed mitogen (PWM)
    c. Lipopolysaccharide (LPS)
    d. Concanavalin A

90. The mitogen that is supplemented to the medium that can act on both B and T lymphocytes is
    a. Phytohaemagglutinin (PHA)
    b. Pokeweed mitogen (PWM)
    c. Lipopolysaccharide (LPS)
    d. Concanavalin A

CHAPTER 3

Animal and Human Cell Lines

Section 1: Animal Cell Lines

1. The animal cell line +/+ SCT represents
   a. Mouse soft and connective tissue
   b. Mouse mast cells
   c. Mouse myeloma cells
   d. Mouse embryo fibroblasts

2. The animal cell line +/+ MGT represents
   a. Mouse soft and connective tissue
   b. Mouse mast cells
   c. Mouse mammary gland cells
   d. Mouse embryo fibroblasts

3. The animal cell line 13762 MAT B III represents
   a. Rat mammary gland cells
   b. Mouse bone marrow
   c. Rat myeloma cells
   d. Mouse embryo fibroblasts

4. The animal cell line 104C1 represents
   a. Rat soft and connective tissue
   b. Rat mammary gland cells
   c. Guinea pig fetal cells
   d. Mouse embryo fibroblasts

5. The animal cell line 127TAg represents
   a. Mouse soft and connective tissue
   b. Mouse mast cells
   c. Mouse myeloma cells
   d. Mouse embryo fibroblasts

6. The animal cell line 121-19B10 represents
   a. Mouse soft and connective tissue
   b. Mouse macrophages
   c. Mouse myeloma cells
   d. Mouse sertoli cells

7. The animal cell line 201-45E9 represents
   a. Mouse soft and connective tissue
   b. Mouse hybridoma cells
   c. Mouse myeloma cells
   d. Mouse pancreatic cells

8. The animal cell line 23 ScCr represents
   a. Mouse macrophage
   b. Mouse kidney cells
   c. Mouse sertoli cells
   d. Mouse embryo fibroblasts

9. The animal cell line 2E8 represents
   a. Mouse soft and connective tissue
   b. Mouse B lymphoblast
   c. Mouse myeloma cells
   d. Mouse embryo fibroblasts

10. The animal cell line 4T1 represents
    a. Mouse pancreatic cancer
    b. Mouse bone cancer
    c. Mouse breast cancer
    d. Mouse skin cancer

11. The animal cell line A7r5 represents
    a. Rat pancreatic cells
    b. Rat fibroblast
    c. Rat tracheal cells
    d. Rat lymphocytes

12. The animal cell line AML12 represents
    a. Mouse hepatocytes
    b. Mouse lymphocytes
    c. Mouse myeloma cells
    d. Mouse epithelial cells

13. The animal cell line ARIP represents
    a. Rat pancreatic tumor
    b. Rat epithelial cells
    c. Rat malignant carcinoma
    d. Rat pituitary tumor

14. The animal cell line B35 represents
    a. Rat pancreatic tumor
    b. Rat epithelial cells
    c. Rat neuroblastoma
    d. Rat pituitary tumor

15. The animal cell line BHK-21 represents
    a. Syrian golden normal kidney cells
    b. Rat transformed epithelial cells
    c. Rat normal neuronal cells
    d. Rat pituitary tumor cells

16. The animal cell line BRL 3A represents
    a. Rat pituitary cells
    b. Rat transformed epithelial cells
    c. Rat liver fibroblasts
    d. Rat normal epithelial cells

17. The animal cell line C6 represents
    a. Rat malignant carcinoma
    b. Rat Glioma
    c. Rat lymphoma
    d. Rat sarcoma

18. The animal cell line CA-77 represents
    a. Rat thyroid carcinoma
    b. Rat Glioma
    c. Rat lymphoma
    d. Rat sarcoma

19. The animal cell line CH1 represents
    a. Murine glioma
    b. Murine lymphoma
    c. Murine carcinoma
    d. Murine sarcoma

20. The animal cell line CCRF S-180 II represents
    a. Murine glioma
    b. Murine lymphoma
    c. Murine carcinoma
    d. Murine sarcoma

21. The animal cell line Cf2Th represents
    a. Murine thymus cells
    b. Canine thymus cells
    c. Guinea pig thymus cells
    d. Golden hamster thymus cells

22. The animal cell line CGBQ represents
    a. Murine fibroblasts
    b. Canine fibroblasts
    c. Goose fibroblasts
    d. Golden fibroblasts

23. The animal cell line CCO represents
    a. Murine fibroblasts
    b. Canine fibroblasts
    c. Goose fibroblasts
    d. Catfish fibroblasts

24. Of the following, the cell line that does not represent Chinese Hamster Ovary cells is
    a. CHO DP-12
    b. CHO INSR 1284
    c. CHO 1-15
    d. CHL/IU

25. Of the following, the cell line that represents normal rat liver epithelial cells is
    a. Clone 9
    b. Clone C
    c. Clone M-3
    d. Clone BC

26. The animal cell line CMMT represents
    a. Rhesus mammary gland
    b. Rhesus salivary gland
    c. Mouse mammary gland

d. Mouse salivary gland

27. The animal cell line Con A – C1 - VICK represents
    a. Rhesus lymphocytes
    b. Rat lymphocytes
    c. Chicken lymphocytes
    d. Mouse lymphocytes

28. The animal cell line CPA 47 represents
    a. Rhesus endothelial cells
    b. Bovine endothelial cells
    c. Chicken endothelial cells
    d. Mouse endothelial cells

29. The animal cell line CRFK represents
    a. Cat endothelial cells
    b. Cat kidney cells
    c. Cat mammary gland cells
    d. Cat smooth muscle ells

30. The animal cell line D-17 represents
    a. Dog osteosarcoma
    b. Mouse sarcoma
    c. Cat glioma
    d. Hamster carcinoma

31. The animal cell line GH4C1 represents
    a. Dog pituitary tumor
    b. Mouse pituitary tumor
    c. Cat pituitary tumor
    d. Rat pituitary tumor

32. The animal cell line GPC-16 represents
    a. Dog pituitary tumor
    b. Guinea pig colorectal adenocarcinoma
    c. Cat pituitary tumor
    d. Rat pituitary tumor

33. The animal cell line GSML represents
    a. Monkey pancreatic cells
    b. Monkey kidney cells

    c. Monkey B lymphocytes
    d. Monkey pituitary tumor

34. The animal cell line H-4-II-E represents
    a. Rat adenocarcinoma
    b. Rat glioma
    c. Rat hepatoma
    d. Rat pituitary tumor

35. The animal cell line Hepa-1c1c7 represents
    a. Mouse adenocarcinoma
    b. Mouse glioma
    c. Mouse hepatoma
    d. Mouse pituitary tumor

36. The animal cell line HIIF-D represents
    a. Chinese hamster ovary cells
    b. Mouse kidney fibroblasts
    c. Monkey pancreatic cells
    d. Rat pituitary cells

37. The animal cell line HIT-T15 represents
    a. Chinese hamster ovary cells
    b. Syrian golden hamster Beta cells of pancreas
    c. Monkey pancreatic cells
    d. Rat pituitary cells

38. The animal cell line 1-11.15 represents
    a. Monkey macrophages
    b. Mouse macrophages
    c. Rat lymphocytes
    d. Rat macrophages

39. The animal cell line JH4 clone 1 represents
    a. Monkey pancreatic fibroblasts
    b. Monkey kidney cells
    c. Guinea pig lung fibroblasts
    d. Monkey pituitary tumor

40. The animal cell line K7M2 wt represents
    a. Mouse osteosarcoma

b. Mouse glioma

c. Mouse adenosarcoma

d. Mouse pituitary tumor

41. The animal cell line KLN 205 represents
    a. Mouse sarcoma
    b. Mouse squamous cell carcinoma
    c. Mouse lymphoma
    d. Mouse leukemia

42. The animal cell line L8 represents
    a. Rat myoblast
    b. Rat sarcoma
    c. Rat glioma
    d. Rat mammary gland tumor

43. The animal cell line LA-4 represents
    a. Mouse osteosarcoma
    b. Mouse glioma
    c. Mouse lung adenoma
    d. Mouse pituitary tumor

44. The animal cell line LLC-RK1 represents
    a. Normal rabbit kidney cells
    b. Rabbit glioma
    c. Rabbit adenosarcoma
    d. Rabbit pituitary tumor

45. The animal cell line MB355 represents
    a. Mouse embryo fibroblast
    b. Mouse glioma
    c. Mouse lung adenoma
    d. Mouse pituitary tumor

46. The animal cell line MC/9 represents
    a. Mouse embryo fibroblast
    b. Mouse liver mast cells
    c. Mouse pancreatic fibroblasts
    d. Mouse kidney fibroblasts

47. The animal cell line RH/K34 represents

a. Dog leukemia
b. Rat leukemia
c. Murine leukemia
d. Rabbit leukemia

48. The animal cell line PIN-14B represents
    a. Rat glioma
    b. Rat pancreatic tumor
    c. Rat lung adenoma
    d. Rat pituitary tumor

49. The animal cell line SBAC represents
    a. Mouse fibroblast
    b. Rabbit fibroblast
    c. Cow fibroblast
    d. Guinea pig fibroblast

50. The animal cell line ST represents
    a. Murine testis cells
    b. Pig testis cells
    c. Rabbit testis cells
    d. Rat testis cells

51. The animal cell line TRA-171 represents
    a. Bat fibroblast
    b. Turtle cardiomyocytes
    c. Mosquito fibroblasts
    d. Murine leydig cells

52. The animal cell line W-20-17 represents
    a. Bat bone marrow stromal cells
    b. Mouse bone marrow stromal cells/fibroblasts
    c. Rat bone marrow stromal cells
    d. Chick bone marrow stromal cells

53. The animal cell line ZF4 represents
    a. Zebra fish embryo cells
    b. Zebra fish liver cells
    c. Zebra fish intestinal cells
    d. Zebra fish epithelial cells

54. The animal cell line ZFL represents
    a. Zebra fish embryo cells
    b. Zebra fish liver epithelial cells
    c. Zebra fish intestinal cells
    d. Zebra fish epithelial cells

55. The animal cell line A-10 represents
    a. Rat aorta cells/myoblasts
    b. Rat liver cells
    c. Rat kidney cells
    d. Rat lymphocytes

56. The animal cell line MDCK represents
    a. Bone marrow stromal cells
    b. Canine kidney epithelial cells
    c. Bovine chondrocytes
    d. Murine trophoblast cells

Section 2: Human Cell Lines

57. The human cell line GH354 represents
    a. Cervical Adenocarcinoma
    b. Sarcoma
    c. Glioma
    d. Normal epithelial cells

58. The human cell line 143B represents
    a. Glioblastoma
    b. Chondrosarcoma
    c. Osteosarcoma
    d. Melanoma

59. The human cell line 184B5 represents
    a. Breast cancer
    b. Normal mammary gland epithelium
    c. Normal corneal epithelium
    d. Lymphoma

60. The human cell line HEK 293T/17 represents
    a. Kidney cells
    b. Mammary epithelium
    c. Lymphocytes

    d. Monocytes

61. The human cell line ED27 represents
    a. Bone marrow stromal cells
    b. Osteocytes
    c. Chondrocytes
    d. Placental trophoblast cells

62. The human cell line C3A represents
    a. Monocytes
    b. Hepatocytes
    c. Osteocytes
    d. Corneal cells

63. The human cell line MCF-7 represents
    a. Lung cancer cells
    b. Prostate cancer cells
    c. Glioma cells
    d. Breast cancer cells

64. The human cell line PC-3 represents
    a. Lung cancer cells
    b. Prostate cancer cells
    c. Glioma cells
    d. Breast cancer cells

65. The human cell line LoVo represents
    a. Non-small cell lung cancer cells
    b. Glioblastoma cells
    c. Colon cancer cells
    d. Bladder cancer cells

66. The human cell line HepG2 represents
    a. Hepatoma cells
    b. Prostate cancer cells
    c. Glioma cells
    d. Breast cancer cells

67. The human cell line OV-MZ-6 represents
    a. Hepatoma cells
    b. Prostate cancer cells

    c. Ovarian cancer cells
    d. Breast cancer cells

68. The human cell line SKOV-3 represents
    a. Hepatoma cells
    b. Prostate cancer cells
    c. Ovarian cancer cells
    d. Breast cancer cells

69. The human cell line HT1080 represents
    a. Fibrosarcoma cells
    b. Glioblastoma cells
    c. Adenocarcinoma cells
    d. Osteosarcoma cells

70. The human cell line A549 represents
    a. Kidney cancer cells
    b. Lung cancer cells
    c. Prostate cancer cells
    d. Osteosarcoma cells

71. The human cell line H358 represents
    a. Kidney cancer cells
    b. Lung cancer cells
    c. Prostate cancer cells
    d. Osteosarcoma cells

72. The human cell line SH-SY5Y represents
    a. Neuroblastoma cells
    b. Osteosarcoma cells
    c. Glioma cells
    d. Myeloma cells

73. The human cell line C3A represents
    a. Hepatocytes
    b. Neurons
    c. Glial cells
    d. Stromal cells

74. The human cell line IMR-32 represents
    a. Neuroblastoma cells

b. Osteosarcoma cells
c. Glioma cells
d. Myeloma cells

75. The human cell line SK-N-AS represents
    a. Melanoma cells
    b. Osteosarcoma cells
    c. Breast cancer cells
    d. Neuroblastoma cells

76. The human cell line SK-N-DZ represents
    a. Melanoma cells
    b. Neuroblastoma cells
    c. Lung adenocarcinoma cells
    d. Myeloma cells

77. The human cell line Huh7 represents
    a. Melanoma cells
    b. Hepatoma cells
    c. Lung adenocarcinoma cells
    d. Breast cancer cells

78. The human cell line L1236 represents
    a. Hodgkin lymphoma cells
    b. Non-Hodgkin lymphoma cells
    c. Sarcoma cells
    d. Adenocarcinoma cells

79. The human cell line U-251 MG represents
    a. Astrocytoma cells
    b. Lymphoma cells
    c. Myeloma cells
    d. Sarcoma cells

80. The human cell line ED27 represents
    a. Fibroblast cells
    b. Stromal cells
    c. Trophoblast cells
    d. Smooth muscle cells

81. The human cell line RT112 represents
    a. Lung adenocarcinoma cells
    b. Bladder carcinoma cells
    c. Neuroblastoma cells
    d. Glioblastoma cells

CHAPTER 4

Hybridoma Technology

1. "Hybridoma" refers to
   a. Hybrid melanoma
   b. Hybrid myeloma
   c. Murine antibodies
   d. Monoclonal antibodies

2. Hybridoma technology is to primarily generate
   a. Monoclonal antibodies
   b. Bispecific antibodies
   c. Hybrid antibodies
   d. Antibody fragments

3. Typically, monoclonal antibodies are largely of the
   a. Murine type
   b. Rabbit antibodies
   c. Human antibodies
   d. Human IgG

4. The most widely used strain of animals for hybridoma technology is the
   a. Wister rats
   b. Albino mouse
   c. Rabbits
   d. Balb/c mice

5. The person other than Cesar Milstein and Georges Kohler who received the Nobel prize for inventing the hybridoma technology was
   a. Edward Jenner
   b. Karl Landsteiner
   c. Niels Jerne
   d. Leonard Herzenberg

6. The term "hybridoma" was coined by
   a. Edward Jenner
   b. Karl Landsteiner
   c. Niels Jerne
   d. Leonard Herzenberg

7. The route of primary immunizations to develop hybridomas are typically
   a. Intramuscular
   b. Intravenous
   c. Intraperitoneal
   d. Subcutaneous

8. The route of final consecutive booster immunizations to develop hybridomas are typically
   a. Intramuscular
   b. Intravenous
   c. Intraperitoneal
   d. Subcutaneous

9. The primary immunizations for hybridoma development typically utilize
   a. Freund's incomplete adjuvant
   b. Freund's complete adjuvant
   c. Alum solution
   d. SDS suspension

10. The secondary immunizations for hybridoma development typically utilize
    a. Freund's incomplete adjuvant
    b. Freund's complete adjuvant
    c. Alum solution
    d. SDS suspension

11. The organ harvested as a source of immunized lymphocytes for hybridoma technology is
    a. Lymph nodes
    b. Bone marrow
    c. Spleen
    d. Payers patches

12. The most commonly used myeloma cell line as a fusion partner for developing murine hybridomas is
    a. Sp2/0

    b. MCF-7
    c. Plasma cells
    d. A549

13. The type of fusogens for somatic cell hybridization that are available does not include
    a. Electroporation
    b. Chemical fusogens such as polyethylene glycol
    c. Sendai virus
    d. UV radiation

14. The most commonly used chemical fusogen for somatic cell hybridization is
    a. Sodium Dodecyl Sulphate
    b. Polyethylene glycol
    c. TEMED
    d. Tween-20

15. The most commonly & safely used % of polyethylene glycol for somatic cell hybridization is
    a. 30
    b. 35
    c. 50
    d. 55

16. The myeloma cells used for hybridoma technology should
    a. Be positive for the HGPRT enzyme
    b. Be negative for the HGPRT enzyme
    c. Should be arrested at the G0 stage
    d. Should be extremely sensitive for the cation of PEG

17. The medium that is used for selecting true hybridomas is
    a. HAT medium
    b. DMEM medium
    c. RPMI medium
    d. Hank's balanced salt solution

18. The selection medium used for hybridoma technology does not necessarily contain
    a. Hypoxanthine
    b. Aminopterine
    c. Thymidine

  d. Alanine

19. The component in the HAT medium that arrests the de novo DNA synthesis pathway is
    a. Hypoxanthine
    b. Aminopterin
    c. Thymidine
    d. Alanine

20. The DNA synthesis pathway that occurs in hybridoma cells is
    a. De novo pathway
    b. Salvage pathway
    c. Purine pathway
    d. Pyramidine pathway

21. The plasma cells that are used for hybridoma technology should be positive for
    a. HGPRT enzyme alone
    b. Ig secretion alone
    c. Both HGPRT and Ig
    d. Neither HGPRT or Ig

22. The dilution method that is used to obtain single clones for the hybridoma technology is
    a. Limiting dilution
    b. Serial dilution
    c. Step-wise dilution
    d. Distribution suspension

23. The most commonly used screening technique to identify positive hybridoma clones is
    a. RIA
    b. ELISA
    c. Double diffusion
    d. Immunoelectrophoresis

24. Hybridomas can grow in the medium
    a. DMEM
    b. RPMI
    c. Both DMEM and RPMI
    d. Hank's Balanced Salt Solution

25. The culture plates that are used for limiting dilution is of the type
    a. 6-well plates
    b. 12-well plates
    c. 24-well plate
    d. 96-well plates

26. The cells present in the culture plate in which limiting dilutions are performed, and those that help stabilize hybridoma clones initially are known as
    a. Feeder cells
    b. Booster cells
    c. Fusion partners
    d. Suspension cells

27. The most preferred or commonly obtained isotype of the monoclonal antibodies is
    a. IgG
    b. IgA
    c. IgM
    d. IgD

28. *In vivo* expansion of hybridoma clones is performed in the
    a. Peritoneal cavity of mice
    b. Serum of rabbits
    c. Embryonated eggs
    d. Serum of mice

29. *In vivo* expansion of hybridoma clones in mice results in the production of
    a. Synovial fluid
    b. Cerebrospinal fluid
    c. Ascetic fluid
    d. Serum

CHAPTER 5

3D Cell Cultures

1. In 3D cell culture, cells
    a. Are exact replicas of *in vivo* conditions
    b. Mimic *in vivo* features
    c. Are more like transplants
    d. Are more like monolayers

2. 3D cell culture techniques render cells
    a. Closer to *in vivo*
    b. Closer to xenotransplants
    c. Closer to explants
    d. Closer to suspension cultures

3. The critical component for obtaining 3D cell cultures is
    a. Matrices or scaffolds
    b. Specific growth factors
    c. Antibiotics
    d. Serum

4. An example of a simple matrix or scaffold to obtain 3D cell cultures is
    a. Silica gel
    b. Fibrinogen
    c. Agarose hydrogel
    d. Starch

5. One feature that 3D cell culture systems provide that is lacking in a 2D culture system is
    a. Presence of other cells
    b. Presence of nutrients
    c. Enhanced cell-to-matrix interactions
    d. Space to grow

6. An example of a natural extracellular matrix to obtain 3D cell cultures is
   a. Decellularized tissue
   b. Hyaluronic acid
   c. Poly lactic-co-glycolic acid
   d. PCL-collagen

7. An example of a natural extracellular matrix to obtain 3D cell cultures does not include
   a. Decellularized tissue
   b. Collagen
   c. Fibrin
   d. Polyethylene glycol

8. An example of a synthetic extracellular matrix to obtain 3D cell cultures is
   a. Gelatine
   b. Fibroin
   c. Polyurethane
   d. Tantalam

9. An example of a synthetic extracellular matrix to obtain 3D cell cultures is
   a. Nitinol
   b. Polycaprolactone
   c. Chitosan
   d. Chitin

10. An example of a biological and synthetic hybrid matrix to obtain 3D cell cultures is
    a. Polyethylene glycol
    b. CNT-ceramic matrix
    c. Hydroxyapatite-collagen hybrid
    d. Chitosan

11. A metal that can be used to prepare matrix to obtain 3D cell cultures is
    a. Magnesium
    b. Silver
    c. Zinc
    d. Steel

12. A metal that can be used to prepare matrix to obtain 3D cell cultures is
    a. Nickel alloy
    b. Thorium

    c. Copper

    d. Zinc

13. Carbon nanotubes as can be used for obtaining 3D cell cultures are typically made of
    a. Graphite
    b. Charcoal
    c. Sulphur
    d. Calcium

14. Laminin that can be used for obtaining 3D cell cultures belongs to the matrix type
    a. Natural extracellular matrix
    b. Synthetic extracellular matrix
    c. Biological and synthetic hybrid matrix
    d. Biodegradable polymer matrix

15. Alginate that can be used for obtaining 3D cell cultures belongs to the matrix type
    a. Natural decellularized tissue matrix
    b. Synthetic extracellular matrix
    c. Biological and synthetic hybrid matrix
    d. Natural extracellular matrix

16. Elastin that can be used for obtaining 3D cell cultures belongs to the matrix type
    a. Natural decellularized tissue matrix
    b. Natural basement membrane matrix
    c. Natural extracellular matrix
    d. Biological and synthetic hybrid matrix

17. Synthetic polymer matrices that can be used to obtain 3D cell cultures do not include
    a. Modified polyethylene glycol forms
    b. Modified hyaluronic acid forms
    c. Gelatine
    d. Polyurethane

18. The type of matrices used for obtaining 3D cell cultures that resemble *in vivo* conditions to the maximum are
    a. Biological and synthetic hybrids

b. Biodegradable synthetic polymers
c. Biodegradable natural matrices
d. Self-assembling protein hydrogels

19. The type of matrices used for obtaining 3D cell cultures that would have maximum batch-to-batch variations are
a. Biological and synthetic hybrids
b. Biodegradable synthetic polymers
c. Biodegradable natural matrices
d. Self-assembling protein hydrogels

20. The type of matrices used for obtaining 3D cell cultures that would have maximum reproducible mechanical and physical properties are
a. Biological and synthetic hybrids
b. Biodegradable synthetic polymers
c. Biodegradable natural matrices
d. Self-assembling protein hydrogels

21. The type of matrices used for obtaining 3D cell cultures that are most compatible to be supplemented with growth factors or other bio-molecules are
a. Biological and synthetic hybrids
b. Biodegradable synthetic polymers
c. Biodegradable natural matrices
d. Self-assembling protein hydrogels

22. The type of matrices used for obtaining 3D cell cultures that have maximum flexibility in terms of mechanical properties are
a. Biological and synthetic hybrids
b. Biodegradable synthetic polymers
c. Biodegradable natural matrices
d. Self-assembling protein hydrogels

23. The type of matrices used for obtaining 3D cell cultures that have maximum porous properties are
a. Biological and synthetic hybrids
b. Biodegradable synthetic polymers
c. Ceramic matrices
d. Self-assembling protein hydrogels

24. The type of matrices used for obtaining 3D cell cultures that have maximum cell-infiltration potential are

  a. Biological and synthetic hybrids
  b. Biodegradable synthetic polymers
  c. Ceramic matrices
  d. Self-assembling protein hydrogels

25. The type of matrices used for obtaining 3D cell cultures that have maximum electric conductivity are
  a. Biological and synthetic hybrids
  b. Biodegradable synthetic polymers
  c. Ceramic matrices
  d. Carbon nanotubes

26. The type of matrices used for obtaining 3D cell cultures that have maximum native tissue integration property are
  a. Biological and synthetic hybrids
  b. Biodegradable synthetic polymers
  c. Natural extracellular matrices
  d. Carbon nanotubes

27. The type of matrices used for obtaining 3D cell cultures that have maximum cell differentiation property are
  a. Biological and synthetic hybrids
  b. Biodegradable synthetic polymers
  c. Natural extracellular matrices
  d. Carbon nanotubes

28. The type of matrices used for obtaining 3D cell cultures that have maximum potential as carriers for drugs, growth factors, *etc.* are
  a. Biological and synthetic hybrids
  b. Biodegradable synthetic polymers
  c. Natural extracellular matrices
  d. Synthetic extracellular matrices

29. The type of matrices used for obtaining 3D cell cultures that have maximum potential for stimulating wound healing are
  a. Biological and synthetic hybrids
  b. Biodegradable synthetic polymers
  c. Natural extracellular matrices
  d. Synthetic extracellular matrices

30. The type of matrices used for obtaining 3D cell cultures that have maximum

potential for studying organ- or disease-specificity are
a. Biological and synthetic hybrids
b. Biodegradable synthetic polymers
c. Natural extracellular matrices
d. Synthetic extracellular matrices

31. The type of matrices used for obtaining 3D cell cultures that have maximum potential for studying chondrocytes, neuronal cells and cardiac tissue constructs are
a. Biological and synthetic hybrids
b. Carbon nanotubes
c. Natural extracellular matrices
d. Synthetic extracellular matrices

32. The morphology of cells grown in 3D conditions will be
a. Spindle shaped and flat
b. Rounded and suspended
c. Polarized and ellipsoid
d. No specific shape

33. 3D culture conditions will enable
a. All cells in culture to be in contact with the medium
b. Nutrient and oxygen gradients
c. Uniform exposure of all cells to supplemented material
d. Synchronous cell cultures

34. 3D culture conditions will enable
a. Enhanced cell junctions
b. Reduced cell junctions
c. Enhanced survival
d. Reduced survival

35. 3D culture conditions will enable
a. All cells in culture to be in a specific cell cycle stage
b. Reduced cell junctions
c. Enhanced cell-to-cell communication and signalling
d. Reduced interactions with the extracellular matrix

36. 3D culture conditions will enhance
a. Increased differentiation
b. Increased cell junctions

    c. Cell-to-cell communication and signalling

    d. All of the above

37. 3D culture conditions will induce
    a. Decreased differentiation
    b. Decreased cell junctions
    c. Decreased drug metabolism
    d. Decreased drug sensitivity

38. 3D culture conditions will induce
    a. Enhanced apoptosis
    b. Enhanced drug resistance
    c. Enhanced potency to cytotoxic drugs
    d. Enhanced susceptibility to anti-proliferative agents

39. 3D cell culture methods include
    a. Scaffold-based methods
    b. Scaffold-free methods
    c. Both scaffold-based and scaffold-free methods
    d. Exclusively hydrogel-based methods

40. The scaffold-free 3D cell culture methods do not include
    a. Hanging-drop method
    b. Those using ultra-low attachment plates
    c. Microfluidic methods
    d. Those using micropatterned surface microplates

41. The scaffold-based 3D cell culture methods do not include
    a. Polymeric hard scaffolds
    b. Those using ultra-low attachment plates
    c. Biologic scaffolds
    d. Those using micropatterned surface microplates

CHAPTER 6

Stem Cells

1. Stem cells are
   a. Differentiated
   b. Undifferentiated
   c. Specialized
   d. Only embryonic

2. Adult stem cells occur at
   a. Inner cell mass
   b. Progenitors
   c. Various tissues
   d. Trophoblast

3. Embryonic stem cells occur at
   a. Inner cell mass
   b. Progenitors
   c. Various tissues
   d. Trophoblast

4. A source of autologous adult stem cells is
   a. Bone marrow
   b. Cord blood
   c. Inner cell mass
   d. Mesoderm

5. The properties of stem cells do not include
   a. Self-renewal
   b. Potency
   c. Differentiation
   d. Meiosis

6. The type of tissue that embryonic stem cells cannot differentiate into is
   a. Endoderm
   b. Mesoderm
   c. Placenta
   d. Ectoderm

7. Totipotent cells are also known as
   a. Pluripotent
   b. Multipotent
   c. Omnipotent
   d. Self-renewable

8. Pluripotent stem cells can generate
   a. All types of cells
   b. Nearly all types of cells
   c. About 3 different types of cells of a lineage
   d. Single type of differentiated cells

9. Totipotent stem cells can generate
   a. All types of cells
   b. Nearly all types of cells
   c. About 3 different types of cells of a lineage
   d. Single type of differentiated cells

10. Multipotent stem cells can generate
    a. All types of cells
    b. Nearly all types of cells
    c. Few types of cells of a particular lineage
    d. Single type of differentiated cells

11. Oligopotent stem cells can generate
    a. All types of cells
    b. Nearly all types of cells
    c. Few types of cells of a particular lineage
    d. Single type of differentiated cells

12. Unipotent stem cells can generate
    a. All types of cells
    b. Nearly all types of cells
    c. Single type of cells and are not capable of self-renewal
    d. Single type of differentiated cells as well as self-renewal property

13. Progenitor cells
    a. Can self-renew
    b. Cannot self-renew
    c. Are differentiated
    d. Are specialised

14. Embryonic stem cells are
    a. Multipotent
    b. Unipotent
    c. Pluripotent
    d. Progenitors

15. The type of tissue that embryonic stem cells cannot differentiate into is
    a. Extra-embryonic membranes
    b. Mesoderm
    c. Endoderm
    d. Ectoderm

16. The major transcription factors that occur in human embryonic stem cells do not include
    a. Sox2
    b. Nanog
    c. Oct-2
    d. Oct-4

17. The major surface antigens that occur in human embryonic stem cells do not include
    a. Stage-specific embryonic antigens 3 & 4
    b. Stage- specific embryonic antigen 3 & 5
    c. Stage- specific embryonic antigen 2 & 3
    d. Stage- specific embryonic antigen 1 & 2

18. Adult stem cells are usually
    a. Pluripotent
    b. Multipotent
    c. Omnipotent
    d. Non self-renewable

19. Adult stem cells are also known as
    a. Omnipotent stem cells
    b. Pluripotent stem cells

     c. Somatic stem cells
     d. Unipotent stem cells

20. Genetic reprogramming can generate
     a. Induced pluripotent stem cells
     b. Embryonic stem cells
     c. Inner cell mass stem cells
     d. Differentiated stem cells

21. Neural stem cells can not generate into
     a. Neurons
     b. Fibroblasts
     c. Astrocytes
     d. Oligodendrocytes

22. The common non-human antigen that human embryonic stem cells express is
     a. Neu5Gc
     b. CD34
     c. Myc
     d. Luc

23. Human embryonic stem cells have high expression of
     a. MHC Class III molecules
     b. MHC Class II molecules
     c. MHC Class I molecules
     d. All of the above

24. The feeder cells on which the human embryonic stem cells have traditionally been cultured are
     a. Human fibroblasts
     b. Mouse embryonic fibroblasts
     c. Rat epithelial cells
     d. Human epithelial cells

25. One of the major identifying marker of hematopoietic stem cells is
     a. Sca-1
     b. CD16
     c. C-Myc
     d. Tie-4

26. Of the following markers, which is not an identification marker for

hematopoietic stem cell marker
a. C-kit
b. CD50
c. C-MPL
d. C4pRq

27. Of the following markers, which is not an identification marker for hematopoietic stem cell marker
a. Tie-2
b. CD150
c. CD50
d. C1qRp

28. A well-defined antigen which is a hematopoietic differentiation antigen is
a. CD38
b. CD48
c. CD8
d. CD244

29. The typical ratio of hematopoietic stem cells relative to other cells of the bone marrow is
a. 1: 10,000
b. 1: 5000
c. 1: 1000
d. 1: 500

CHAPTER 7

History of Microbiology

1. The first recorded observation and description of microbes was performed by
   a. Antony van Leeuwenhoek
   b. Fracastorius
   c. Von Plenciz
   d. Oliver Wendell

2. The *Bacillus anthracis* were first observed in 1850 by
   a. Davine and Holmes
   b. Augustino Bassi
   c. Davine and Pollender
   d. Louis Pasteur

3. The first scientific study on fermentation was carried out by
   a. Antony van Leeuwenhoek
   b. Von Plenciz
   c. Louis Pasteur
   d. Peyton Rous

4. The abiogenesis generation of microbes was put forth by
   a. Spallanzani
   b. Needham
   c. Hansen
   d. Yersin

5. The germ theory microbes was proposed by
   a. Spallanzani
   b. Needham
   c. Pasteur
   d. Yersin

6. Attenuation was first accidentally observed in
   a. Chicken cholera bacilli
   b. Anthrax
   c. Mycobacterium
   d. Small pox virus

7. The term 'vaccine' was coined by
   a. Louis Pasteur
   b. Edward Jenner
   c. Robert Koch
   d. Peyton Rous

8. The first recorded scientific antiseptic methods were practiced by
   a. Pasteur
   b. Walter Reed
   c. Lister
   d. Neisser

9. The first documented & elaborated scientific studies on bacteriology were performed by
   a. Edward Jenner
   b. Peyton Rous
   c. Walter Reed
   d. Robert Koch

10. The first bacterial staining technique was reported by
    a. Walter Reed
    b. Robert Koch
    c. Metchnikoff
    d. Till Domagk

11. The first method of obtaining pure cultures of bacteria was demonstrated by
    a. Walter Reed
    b. Robert Koch
    c. Metchnikoff
    d. Till Domagk

12. The tuberculosis bacteria was discovered in the year
    a. 1882
    b. 1769
    c. 1980

d. 1892

13. The cholera vibrio was discovered in the year
    a. 1782
    b. 1883
    c. 1980
    d. 1892

14. The leprosy bacillus was first described by
    a. Hansen
    b. Ehrlich
    c. Burnet
    d. Ivanovsky

15. The gonococcus was first described by
    a. Hansen
    b. Ehrlich
    c. Neisser
    d. Ivanovsky

16. The staphylococcus was first described by
    a. Hansen
    b. Ehrlich
    c. Burnet
    d. Ogston

17. The diphtheria bacillus was first described by
    a. Loeffler
    b. Ehrlich
    c. Burnet
    d. Ivanovsky

18. The tetanus bacillus was first described by
    a. Hansen
    b. Ehrlich
    c. Nicolaier
    d. Ivanovsky

19. The pneumococcus was first described by
    a. Nicolaier
    b. Fraenkel

    c. Loeffler

    d. Ogston

20. The causative pathogen of malta fever was first described by
    a. Hansen
    b. Bruce
    c. Burnet
    d. Ivanovsky

21. The syphilis spirochete was first described by
    a. Hansen and Burnet
    b. Ehrlich and Ogston
    c. Shaudinn and Hoffmann
    d. Ivanovsky and Nicolaier

22. The toxin induced pathogenesis was first explained by
    a. Roux and Yersin
    b. Shaudinn and Hoffmann
    c. Ogston and Nicolaier
    d. Hansen and Ivanovsky

23. The concept of antitoxins that work against toxins was first explained by
    a. Hansen
    b. Pasteur
    c. Jenner
    d. Ehrlich

24. The postulates that help in describing a pathogenic, disease causative agent was first put forth by
    a. Koch
    b. Jenner
    c. Ehrlich
    d. Hoffmann

25. The mosaic disease in plants was first described by
    a. Ivanovsky
    b. Ogston
    c. Ehrlich
    d. Hoffmann

26. The term 'virus' was first coined by

    a. Beijerinck
    b. Hoffmann
    c. Ivanovsky
    d. Pasteur

27. The yellow fever was first scientifically studied by
    a. Goodpasture
    b. Peyton Rous
    c. Walter Reed
    d. Walter Sutton

28. The poliomyelitis was first scientifically studied by
    a. Goodpasture
    b. Landsteiner and Popper
    c. Walter Reed
    d. Walter Sutton and Ivanovsky

29. The electron microscope was invented by
    a. Ruska
    b. Rous
    c. Sutton
    d. Hoffmann

30. Culture of viruses in developing chick embryos was first introduced by
    a. Ruska
    b. Koch
    c. Goodpasture
    d. Ellerman

31. Penicillin was discovered by
    a. Fleming
    b. Burnet
    c. Metchnikoff
    d. Nuttall

32. The microscopic observations of the mold fruiting bodies were first documented by
    a. Robert Hooke
    b. Anton von Leeuwenhoek
    c. Ferdinand Cohn
    d. Martinus Beijerinck

33. The first taxonomic classification of bacteria was proposed by
    a. Louis Pasteur
    b. Robert Koch
    c. Ferdinand Cohn
    d. Sergei Winogradsky

34. The person considered the 'father of medical microbiology' is
    a. Louis Pasteur
    b. Robert Koch
    c. Ferdinand Cohn
    d. Sergei Winogradsky

35. The person considered the 'father of microbiology' is
    a. Louis Pasteur
    b. Robert Koch
    c. Ferdinand Cohn
    d. Sergei Winogradsky

36. The spontaneous generation theory was disproved by
    a. Joseph Lister
    b. Louis Pasteur
    c. Robert Koch
    d. Antony von Leeuwenhoek

37. The discovery that viruses require a living cell for multiplication was made by
    a. Alexander Fleming
    b. Paul Ehrlich
    c. Martinus Beijerinck
    d. Dmitri Ivanovsky

38. The terms "aerobic" and "anaerobic" were first introduced by
    a. Louis Pasteur
    b. Thomas Huxley
    c. James Strick
    d. Ferdinand Cohn

39. The theory of biogenesis was put forth by
    a. James Strick
    b. Robert Koch
    c. Thomas Huxley
    d. John Tyndall

40. The earliest of bacterial classification and the introduction of the term "Bacillus" were by
    a. John Tyndall
    b. Louis Pasteur
    c. Robert Koch
    d. Ferdinand Cohn

41. The association of *Mycobacterium leprae* as the causative agent of leprosy was first identified by
    a. Edward Jenner
    b. Armauer Hansen
    c. Robert Koch
    d. Thomas Burrill

42. The association of a bacteria as a causative agent of anthrax was first identified by
    a. Edward Jenner
    b. Armauer Hansen
    c. Robert Koch
    d. Thomas Burrill

43. The first scientific documentation of bacteria using cover slips and staining was performed by
    a. Edward Jenner
    b. Armauer Hansen
    c. Robert Koch
    d. Thomas Burrill

44. The first scientific documentation fractional sterilization of bacteria was performed by
    a. John Tyndall
    b. Armauer Hansen
    c. Robert Koch
    d. Thomas Burrill

45. Attenuation of a virulent pathogen towards vaccine generation was first described by
    a. Edward Jenner
    b. Louis Pasteur
    c. Robert Koch
    d. Paul Ehrlich

46. The utilization of methylene blue as a bacterial stain was first described by
    a. Edward Jenner
    b. Louis Pasteur
    c. Robert Koch
    d. Paul Ehrlich

47. The utilization of chemical sterilization for surgical applications was first described by
    a. Edward Jenner
    b. Louis Pasteur
    c. Robert Koch
    d. Paul Ehrlich

48. The utilization of agar for culturing bacteria was first performed by
    a. Angelina Fannie and Walther Hesse
    b. Timothy Lewis and Griffith Evans
    c. Thomas Burill and Joseph Lister
    d. Robert Koch and John Tyndall

49. The discovery of first bacteriophage was made by
    a. Chaim Weizmann
    b. Frederick Twort
    c. Paul Ehrlich
    d. Sigurd Orla-Jensen

50. The term "bacteriophage" was coined by
    a. Chaim Weizmann
    b. Frederick Twort
    c. Felix d'Herelle
    d. Paul Ehrlich

51. The utilization of blood agar for culturing bacteria was first performed by
    a. Angelina Fannie
    b. Walther Hesse
    c. James Brown
    d. Timothy Lewis

52. The person who first distinguished viruses and bacteria was
    a. Thomas Rivers
    b. Albert Jan Kluyver
    c. Everitt Murray

    d. Frederick Griffith

53. Bacterial transformation was discovered by
    a. Thomas Rivers
    b. Albert Jan Kluyver
    c. Everitt Murray
    d. Frederick Griffith

54. Penicillin was discovered by
    a. Alexander Fleming
    b. Frederick Griffith
    c. Paul Ehrlich
    d. Louis Pasteur

55. Viral cultures in eggs was first described by
    a. Angelina Fannie and Walther Hesse
    b. Timothy Lewis and Griffith Evans
    c. Alice Woodruff and Ernest Goodpasture
    d. Robert Koch and John Tyndall

56. The first electron micrograph of bacteria was published by
    a. Alice Evans
    b. William de Monbreun
    c. Rebeccca Lancefield
    d. Ladislaus Laszlo Marton

57. The first electron micrograph of a virus was published by
    a. Alice Evans
    b. Ladislaus Laszlo Marton
    c. Helmuth Ruska
    d. William de Monbreun

58. Bacterial conjugation was first described by
    a. Robert Koch and John Tyndall
    b. Joshua Lederberg and Edward Tatum
    c. John H Weller and Frederick C Robins
    d. Michael Heidelberger and Colin Macleod

59. The term "plasmid" was coined by
    a. Chaim Weizmann
    b. Frederick Twort

    c. Joshua Lederberg
    d. Felix d'Herrelle

60. Transduction by bacteriophages was first described by
    a. Robert Koch and John Tyndall
    b. Joshua Lederberg and Norton Zinder
    c. John H Weller and Frederick C Robins
    d. Michael Heidelberger and Colin Macleod

61. That the bacterial DNA was circular was discovered by
    a. Robert Koch and John Tyndall
    b. Joshua Lederberg and Norton Zinder
    c. Francois Jacob and Elie Wollman
    d. Michael Heidelberger and Colin Macleod

62. Transfer of antibiotic resistance in bacteria was discovered by
    a. John Tyndall
    b. Sawada O
    c. Norton Zinder
    d. Colin Macleod

63. Viriods were discovered by
    a. Theodor O. Diener
    b. John Tyndall
    c. Norton Zinder
    d. Colin Macleod

64. Reverse transcriptase was first described by
    a. Robert Koch and John Tyndall
    b. Joshua Lederberg and Edward Tatum
    c. Howard Temin and David Baltimore
    d. Michael Heidelberger and Colin Macleod

65. Smallpox was officially declared as eradicated in the year
    a. 1979
    b. 1969
    c. 1953
    d. 1956

66. Human Immunodeficiency Virus (HIV) was first described by
    a. Robert Koch and John Tyndall

   b. Luc Montagnier and Robert Gallo
   c. Howard Temin and David Baltimore
   d. Michael Heidelberger and Colin Macleod

67. The microbe whose genome was first completely sequenced was
   a. *Escherichia coli*
   b. *Haemophilus influenza*
   c. *Neisseria gonorrhoeae*
   d. *Tobacco Mosaic Virus*

CHAPTER 8

Microbial Taxonomy and Diversity

1. Numerical taxonomy for the classification of bacteria relies on
   a. Genotypic features
   b. Phenotypic features
   c. Genetic make-up
   d. Metabolism type

2. Phenotypic manifestations that can be used to classify bacteria do not include
   a. Gram staining susceptibility
   b. Nutrient requirements
   c. Size of genome
   d. Morphology

3. Different strains are differ in
   a. Descendance
   b. Major phenotypic attributes
   c. Surface antigens
   d. Major genotypic attributes

4. The relationships among microbes can be represented as
   a. Dendrograms
   b. Dot-plots
   c. Evolutionary trees
   d. Cladodes

5. The scientific approach for life form classification was introduced by
   a. Carl von Linne
   b. Charles Darwin
   c. Gregor Mendel
   d. Antonie van Leeuwenhoek

6. The trait/characteristic that is used for microbial classification does not include
   a. Morphology
   b. Lack of true nucleus
   c. Phylogeny
   d. Metabolism

7. The earlier classification term of microbes "Moneres" was redesignated as "Prokaryotes" by
   a. Haeckel
   b. Cohn
   c. Chatton
   d. Jensen

8. The group to which Gram-negative bacteria belong is
   a. Mendocutes
   b. Firmicutes
   c. Mollicutes
   d. Gracilicutes

9. The group to which Gram-positive bacteria belong is
   a. Mendocutes
   b. Firmicutes
   c. Mollicutes
   d. Gracilicutes

10. The suffix –ia represents the microbial classification level of
    a. Class
    b. Order
    c. Family
    d. Subfamily

11. The suffix –idae represents the microbial classification level of
    a. Class
    b. Subclass
    c. Family
    d. Subfamily

12. The suffix –ales represents the microbial classification level of
    a. Family
    b. Subfamily

    c. Order

    d. Suborder

13. The suffix –ineae represents the microbial classification level of
    a. Family
    b. Subfamily
    c. Order
    d. Suborder

14. The suffix –aceae represents the microbial classification level of
    a. Family
    b. Subfamily
    c. Order
    d. Suborder

15. The suffix –oideae represents the microbial classification level of
    a. Family
    b. Subfamily
    c. Order
    d. Suborder

16. The correct order of taxonomic hierarchy is
    a. Kingdom, Class, Order, Family, Genus, Species
    b. Kingdom, Order, Class, Family, Genus, Species
    c. Kingdom, Class, Family, Order, Genus, Species
    d. Kingdom, Family, Order, Class, Genus, Species

17. The Biovars represent
    a. Strains with different physiology and biochemical properties
    b. Strains that re different only morphologically
    c. Strains with different antigen profiles
    d. Strains that differ only physiologically

18. The Serovars represent
    a. Strains with different physiology and biochemical properties
    b. Strains that re different only morphologically
    c. Strains with different antigen profiles
    d. Strains that differ only physiologically

19. The classification type which considers several biological traits is the
    a. Numerical taxonomy

b. Phylogenetic system
c. Natural classification
d. Phenetic system

20. The classification type which considers the overall similarity of organisms is
    a. Numerical taxonomy
    b. Phylogenetic system
    c. Natural classification
    d. Phenetic system

21. The classification type which considers an evolutionary basis is
    a. Numerical taxonomy
    b. Phylogenetic system
    c. Natural classification
    d. Phonetic system

22. The phylogenetic type of classification is also known as
    a. Phyletic taxonomy
    b. Phenetic system
    c. Natural classification
    d. Phonetic system

23. The type of classification that was predominantly followed in the 1st Edition of *Bergey's Manual of Systematic Bacteriology* is
    a. Phyletic taxonomy
    b. Phenetic system
    c. Natural classification
    d. Phonetic system

24. The type of classification that was predominantly followed in the 2nd Edition of *Bergey's Manual of Systematic Bacteriology* is
    a. Phyletic taxonomy
    b. Phenetic system
    c. Phylogenetic system
    d. Phonetic system

25. New pathogens that are difficult to culture can be identified by
    a. 16SrRNA
    b. RNA
    c. DNA
    d. dsDNA

26. The Adansonian classification of bacteria considers
    a. Weighted characteristics
    b. An individual clone
    c. The strain differences
    d. All characteristics that are expressed while the study is being undertaken

27. The Adansonian classification of bacteria is also known as
    a. Phyletic taxonomy
    b. Phenetic system
    c. Phylogenetic system
    d. Phonetic system

28. The branch of bacterial taxonomy that utilizes computational analysis of large data for classification is
    a. Numerical taxonomy
    b. Phylogenetic system
    c. Natural classification
    d. Phonetic system

29. Classification of bacteria based on their biochemical properties result in obtaining
    a. Biotypes
    b. Serotypes
    c. Phage types
    d. Colicin types

30. Classification of bacteria based on their antigenic properties result in obtaining
    a. Biotypes
    b. Serotypes
    c. Phage types
    d. Colicin types

31. Classification of bacteria based on their bacteriophage susceptibility result in obtaining
    a. Biotypes
    b. Serotypes
    c. Phage types
    d. Colicin types

32. Classification of bacteria based on their bacteriocin production properties result in obtaining

a. Biotypes
b. Serotypes
c. Phage types
d. Colicin types

33. The generic name of bacteria in scientific nomenclature is usually a
    a. Latin noun
    b. Adjective
    c. Common meaning
    d. Species property

34. The specific epithet of bacteria in scientific nomenclature is usually not a
    a. Latin noun
    b. An animal where is can commonly occur
    c. A disease that can be caused by the species
    d. The geographical location from where it was first identified

35. In the taxonomic classification of viruses, the suffix –viridae is at the level of
    a. Species
    b. Genus
    c. Family
    d. Superfamily

36. The number of classes of viruses in accordance to the Baltimore classification is
    a. 5
    b. 6
    c. 7
    d. 8

37. The group that double-stranded DNA viruses belong to in accordance to the Baltimore classification is
    a. Group I
    b. Group II
    c. Group III
    d. Group IV

38. The group that single-stranded DNA viruses belong to in accordance to the Baltimore classification is
    a. Group I
    b. Group II

    c. Group III
    d. Group IV

39. The group that double-stranded RNA viruses belong to in accordance to the Baltimore classification is
    a. Group I
    b. Group II
    c. Group III
    d. Group IV

40. The group that positive-sense single-stranded RNA viruses belong to in accordance to the Baltimore classification is
    a. Group IV
    b. Group V
    c. Group VI
    d. Group VII

41. The group that negative-sense single-stranded RNA viruses belong to in accordance to the Baltimore classification is
    a. Group IV
    b. Group V
    c. Group VI
    d. Group VII

42. The group that reverse transcribing diploid single-stranded RNA viruses belong to in accordance to the Baltimore classification is
    a. Group IV
    b. Group V
    c. Group VI
    d. Group VII

43. The group that reverse transcribing circular double-stranded RNA viruses belong to in accordance to the Baltimore classification is
    a. Group IV
    b. Group V
    c. Group VI
    d. Group III

44. The suffix –virales is used at the viral classification level of
    a. Order
    b. Family

    c. Subfamily

    d. Genus

45. The suffix –viridae is used at the viral classification level of
    a. Order
    b. Family
    c. Subfamily
    d. Genus

46. The suffix –virinae is used at the viral classification level of
    a. Order
    b. Family
    c. Subfamily
    d. Genus

47. The suffix –virus is used at the viral classification level of
    a. Family
    b. Genus
    c. Species
    d. Genus and species

48. The spherical bacteria are
    a. Cocci
    b. Bacilli
    c. Vibrio
    d. Spirochete

49. The rodlike bacteria are
    a. Cocci
    b. Bacilli
    c. Vibrio
    d. Spirochete

50. The curve-shaped bacteria do not include
    a. Vibrio
    b. Spirillum
    c. Spirochete
    d. Bacillus

51. Random clumps are formed by
    a. Streptococci

    b. Staphylococci
    c. Pneumococci
    d. Spirochetes

52. Long strand of thread-like structures are formed by
    a. Streptococci
    b. Staphylococci
    c. Pneumococci
    d. Spirochetes

53. The rod-shaped (Bacilli) bacteria normally occur as
    a. Singe cells
    b. Pairs
    c. Clumps
    d. Chains

54. The type of bacteria that can form square or cubical packets are
    a. Cocci
    b. Bacilli
    c. Spirillum
    d. Spirochetes

55. The type of bacteria that can form a helical structure are
    a. Cocci
    b. Bacilli
    c. Spirillum
    d. Spirochetes

56. The microbes that can tolerate high temperatures are known as
    a. Thermophiles
    b. Halophiles
    c. Methanogens
    d. Aerotolerant

57. The microbes that can tolerate high salt concentrations are known as
    a. Thermophiles
    b. Halophiles
    c. Methanogens
    d. Aerotolerant

58. The group of bacteria which do not require oxygen, but can survive in its

presence are known as
   a. Thermophiles
   b. Halophiles
   c. Methanogens
   d. Aerotolerant

59. The group of bacteria for which the presence of oxygen does not affect survival are known as
   a. Obligate anaerobes
   b. Obligate aerobes
   c. Facultative aerobes
   d. Aerotolerant

60. The group of bacteria for which the presence of oxygen is detrimental for survival are known as
   a. Obligate anaerobes
   b. Obligate aerobes
   c. Facultative aerobes
   d. Aerotolerant

61. The group of bacteria for which the presence of oxygen is essential for survival are known as
   a. Obligate anaerobes
   b. Obligate aerobes
   c. Facultative aerobes
   d. Aerotolerant

62. The group of bacteria which use NH_3 or H_2S for ATP production in the absence of sunlight are known as
   a. Photoheterotrophs
   b. Photoautotrophs
   c. Chemoheterotrophs
   d. Chemolithotrophs

63. The group of bacteria which use H_2S for ATP production in the presence of sunlight are known as
   a. Photoheterotrophs
   b. Photoautotrophs
   c. Chemoheterotrophs
   d. Chemolithotrophs

64. The group of bacteria which use sunlight for ATP production but do not synthesize own carbon molecules are known as
    a. Photoheterotrophs
    b. Photoautotrophs
    c. Chemoheterotrophs
    d. Chemolithotrophs

65. The group of bacteria which use complex organic molecules as sources of carbon and energy are known as
    a. Photoheterotrophs
    b. Photoautotrophs
    c. Chemoheterotrophs
    d. Chemolithotrophs

CHAPTER 9

Microbial Structure and Function

1. The dense coat of bacteria, if present is called
   a. Cell wall
   b. Capsule
   c. Plasma membrane
   d. Envelop

2. Loosely associated external bacterial coat is called
   a. Cell wall
   b. Capsule
   c. Slime layer
   d. Envelop

3. Bacterial capsules and slime layers are composed of
   a. Proteins and polysaccharides
   b. Only proteins
   c. Only polysaccharides
   d. Neither proteins nor polysaccharides

4. Bacterial strains that are referred to as "smooth" contain
   a. Capsules
   b. Cell walls
   c. Envelop
   d. Slime layer

5. Bacterial strains that are referred to as "smooth" do not contain
   a. Capsules
   b. Cell walls
   c. Envelop
   d. Slime layer

6. Capsules function for
   a. Staining bacteria
   b. Increasing the pathogenicity of bacteria
   c. Identifying bacteria
   d. Culturing bacteria

7. Cell walls of bacteria are also known as
   a. Capsules
   b. Membrane
   c. Envelop
   d. Slime layer

8. The most abundant bacterial cell wall component is
   a. Peptidoglycan
   b. Dextrose
   c. Starch
   d. Cellulose

9. The number of peptidoglycan layers that Gram-positive bacteria have is
   a. 10-50
   b. 5-10
   c. 1-5
   d. 4

10. The number of peptidoglycan layers that Gram-negative bacteria have is
    a. 1-4
    b. 2-10
    c. 1-2
    d. 5-10

11. The shape of Gram-positive bacteria is
    a. Rigid
    b. Flexible
    c. Can be rigid or flexible
    d. No definite shape

12. The shape of Gram-negative bacteria is
    a. Rigid
    b. Flexible
    c. Can be rigid or flexible
    d. No definite shape

13. The Gram-positive bacteria are usually
    a. Rods only
    b. Cocci only
    c. Rods and cocci
    d. Spirals

14. The Gram-negative bacteria are usually
    a. Rods
    b. Rods and cocci
    c. Rods, cocci and spirals
    d. Rods, cocci, spirals and pleomorphic

15. Spore formation occurs in
    a. Gram-positive bacteria
    b. Gram-negative bacteria
    c. Both Gram-positive and –negative bacteria
    d. The Gram-staining does not matter

16. Spore formation does not commonly occur in
    a. Gram-positive bacteria
    b. Gram-negative bacteria
    c. Both Gram-positive and –negative bacteria
    d. The Gram-staining does not matter

17. Penicillin has a greater inhibitory potential on
    a. Gram-positive bacteria
    b. Gram-negative bacteria
    c. Both Gram-positive and –negative bacteria
    d. The Gram-staining does not matter

18. Aniline dyes have a greater inhibitory potential on
    a. Gram-positive bacteria
    b. Gram-negative bacteria
    c. Both Gram-positive and –negative bacteria
    d. The Gram-staining does not matter

19. Protoplasts can be obtained from
    a. Gram-positive bacteria
    b. Gram-negative bacteria
    c. Both Gram-positive and –negative bacteria
    d. The Gram-staining does not matter

20. Spheroplasts can be obtained from
    a. Gram-positive bacteria
    b. Gram-negative bacteria
    c. Both Gram-positive and –negative bacteria
    d. The Gram-staining does not matter

21. Exoenzymes are directly released to the exterior by
    a. Gram-positive bacteria
    b. Gram-negative bacteria
    c. Both Gram-positive and –negative bacteria
    d. The Gram-staining does not matter

22. Exoenzymes are retained in the periplasmic space by
    a. Gram-positive bacteria
    b. Gram-negative bacteria
    c. Both Gram-positive and –negative bacteria
    d. The Gram-staining does not matter

23. Endotoxins are produced largely by
    a. Gram-positive bacteria
    b. Gram-negative bacteria
    c. Both Gram-positive and –negative bacteria
    d. The Gram-staining does not matter

24. Exotoxins are produced largely by
    a. Gram-positive bacteria
    b. Gram-negative bacteria
    c. Both Gram-positive and –negative bacteria
    d. The Gram-staining does not matter

25. The average length of *Escherichia coli* is
    a. 1 μm
    b. 2 μm
    c. 3 μm
    d. 4 μm

26. The average diameter of *Escherichia coli* is
    a. 0.5 μm
    b. 1.0 μm
    c. 1.5 μm
    d. 2.0 μm

27. The cell envelope of bacteria comprises of
    a. Plasma membrane
    b. Cell wall
    c. Plasma membrane and cell wall
    d. Capsule and cell wall

28. The plasma membrane components of bacteria which is analogous with sterols of eukaryotic cells are
    a. Hopanoids
    b. Fatty acids
    c. Cyclic groups
    d. Glycerol teichoic acids

29. The region between bacterial outer membrane and the cytoplasm is
    a. Lipid bilayer
    b. Periplasm
    c. Capsule
    d. Envelope

30. Signalling proteins of bacteria are known to be imbedded in the
    a. Lipid bilayer
    b. Periplasm
    c. Capsule
    d. Envelope

31. Bacterial fimbriae are also referred to as
    a. Cilia
    b. S-layer
    c. Attachment pili
    d. Sex pili

32. Bacterial fimbriae are tubes made of
    a. Peptidoglycan
    b. Glycolipids
    c. Phospholipids
    d. Proteins

33. Bacterial fimbriae are generally
    a. Located on the entire surface
    b. Located at one pole
    c. Located at either poles

d. Present as groups or tufts

34. Bacterial fimbriae function for
    a. Attachment
    b. Motility
    c. Attachment and motility
    d. Conjugation

35. Bacterial pili mostly function for
    a. Attachment
    b. Motility
    c. Attachment and motility
    d. Conjugation

36. The type of bacterial pili that help in attachment to surfaces is
    a. Type II
    b. Type III
    c. Type IV
    d. Type V

37. Pili are present in
    a. Gram-negative bacteria
    b. Gram-positive bacteria
    c. Both Gram-negative and –positive bacteria
    d. Only during conjugation

38. The polymers that are secreted by bacteria to the exterior are known as
    a. Peptidoglycan
    b. Slayer
    c. Glycocalyx
    d. M-protein

39. The flagella in lophotrichous bacteria are located
    a. As a single flagellum
    b. As a group/tuft at one end
    c. As a group/tuft at both ends
    d. All around the bacterial surface

40. The flagella in monotrichous bacteria are located
    a. As a single flagellum
    b. As one group/tuft at one end

    c. As a group/tuft at both ends
    d. All around the bacterial surface

41. The flagella in amphitrichous bacteria are located
    a. As a single flagellum at the two poles
    b. As a group/tuft at one end
    c. As a group/tuft at both ends
    d. All around the bacterial surface

42. The flagella in peritrichous bacteria are located
    a. As a single flagellum
    b. As a group/tuft at one end
    c. As a group/tuft at both ends
    d. All around the bacterial surface

43. The bacterial flagellum structure does not comprise of
    a. Filament
    b. Mitochondria
    c. Motor complex
    d. Hook

44. The bacterial DNA is usually
    a. Linier
    b. Singe stranded
    c. Circular
    d. Chromatin-like

45. The extra-chromosomal DNA of bacteria are the
    a. Introns
    b. Exons
    c. Nucleoid
    d. Plasmids

46. The most abundant bacterial intracellular structure is
    a. Plasmids
    b. Ribosomes
    c. Nucleoids
    d. Endoplasmic reticulum

47. The bacterial ribosome type is
    a. 70S

    b. 80S
    c. 60S
    d. 40S

48. The 70S ribosome of bacteria is composed of
    a. 50S and 30S subtypes
    b. 40S and 30S subtypes
    c. 50S and 20S subtypes
    d. 40S and 20S subtypes

49. The 50S ribosomal subunit of bacteria is composed of
    a. 20S and 5S rRNA
    b. 22S and 5S rRNA
    c. **23S and 5S** rRNA
    d. 25S and 5S rRNA

50. The 30S ribosomal subunit of bacteria is composed of
    a. 20S rRNA
    b. 19S rRNA
    c. 16S rRNA
    d. 5S rRNA

51. The bacterial intracellular membrane that functions for photosynthesis is
    a. Plasma membrane
    b. Cell wall
    c. Glycocalyx
    d. Chromatophores

52. Inorganic phosphate complexed inclusions that are found in the bacterial cytoplasm are
    a. Volutin granules
    b. Chromatic granules
    c. Sulphur granules
    d. Polyhydroxyalkanoates

53. Volutin granules that are found in bacterial cytoplasm are also known as
    a. Sulphur granules
    b. Polyhydroxyalkanoates
    c. Metachromatic granules
    d. Inclusions

54. Micro-compartments occurring within bacterial cells typically store
    a. Gases
    b. Carbohydrates
    c. Enzymes
    d. Pigments

55. The intracellular structures of bacteria referred to as "polyhedral organelles" are
    a. Vesicles
    b. Vacuoles
    c. Micro-compartments
    d. Granules

56. The component of bacterial endospores that is known to confer heat resistance is
    a. Dipicolinic acid
    b. Magnetite
    c. Greigite
    d. Dextran

57. The net negative charge of the cell walls of Gram-positive bacteria is provided by
    a. Dipicolinic acid
    b. Peptidoglycan
    c. Teichoic acid bonds
    d. Lipids

58. The S-layer is typical of
    a. Gram-positive bacteria
    b. Gram negative bacteria
    c. Both Gram-positive and –negative bacteria
    d. Gram staining does not matter

59. The bacteria capsule is composed of
    a. Proteins
    b. Glycoproteins
    c. Polysaccharides
    d. Glycolipids

60. The bacterial capsule is located
    a. Outside the S-layer

b. Inside the S-layer
c. Within the cytoplasmic membrane
d. Within the envelope

61. The net negative charge of the cell walls of Gram-negative bacteria is provided by
    a. Dipicolinic acid
    b. Peptidoglycan
    c. Teichoic acid bonds
    d. Lipopolysaccharides

62. The antigenic properties of bacterial strains is conferred by the cell wall component
    a. Dipicolinic acid
    b. Peptidoglycan
    c. Teichoic acid bonds
    d. Lipopolysaccharides

63. The passive transport of small molecules across bacterial outer membrane is possible due to
    a. Transporter proteins
    b. Porins
    c. Transducing proteins
    d. Phosphodiesters

64. Pseudoperiplasm occurs in
    a. Gram-positive bacteria
    b. Gram-negative bacteria
    c. Mycobacteria
    d. Both Gram-positive and –negative bacteria

65. Bacterial endotoxins are composed of
    a. Polysaccharides
    b. Lipopolysaccharides
    c. Lipid A
    d. Lipopolysaccharides and Lipid A

66. The functions of the bacterial S-layer do not include
    a. Low pH resistance
    b. Adhesion
    c. Conjugation

    d. Membrane stabilization

67. The structure of bacterial cell that confers antibiotic resistance is
    a. Nucleoid
    b. Plasmid
    c. Vacuoles
    d. Peptidoglycans

68. The cytoskeletal structural element of prokaryotes that forms a ring-like constriction during cell division is
    a. Actin
    b. Myosin
    c. FtsZ
    d. Tubulin

69. The cytoskeletal structural element important for the maintenance of shape in non-spherical bacteria is
    a. FTsZ
    b. MreB
    c. Crescentin
    d. Actin

70. The cytoskeletal structural element of bacteria that is analogous with the intermediate filaments of eukaryotes is
    a. FTsZ
    b. MreB
    c. Crescentin
    d. Actin

71. Plasmid separation during bacterial division is facilitated by the structural element
    a. ParM
    b. FTsZ
    c. MreB
    d. MinCDE system

72. The positioning of the septum during bacterial division is maintained by the structural element
    a. ParM
    b. FTsZ
    c. MreB

    d. MinCDE system

73. The filaments that are found throughout the rod-shaped proteobacterium are
    a. ParM
    b. Bactofilin
    c. Crenactin
    d. Cresentin

74. The infectious viral structure is
    a. Capsid
    b. DNA
    c. RNA
    d. Virion

75. The viral capsid is predominantly composed of
    a. Protein
    b. Glycoprotein
    c. Phospholipids
    d. Nucleic acids

76. The viral structure that combines the capsid and nucleic acid is known as the
    a. Virion
    b. Capsid
    c. Nucleocaspid
    d. Nucleoid

77. The viral capsid shape does not include the structural shape
    a. Helical
    b. Spherical
    c. Icosahedral
    d. Quasispherical

78. The number of identical units that comprise of the icosahedral viral capsid is
    a. 10
    b. 15
    c. 20
    d. 25

79. The number of proteins that the picornavirus nucleocapsid structure is made of is
    a. 2

b. 4

c. 6

d. 8

80. The protein type that is located in the interior of the picornavirus nucleocapsid structure is
    a. VP1
    b. VP2
    c. VP3
    d. VP4

81. The membrane external to the nucleocapsid in some viruses is known as
    a. Envelope
    b. Capsule
    c. Capsid
    d. Coat

82. The viral envelope is largely composed of
    a. Single layer of phospholipids
    b. Double layer of phospholipids
    c. Single layer of glycolipids
    d. Double layer of glycolipids

CHAPTER 10

Microbial Growth and Cultures

1. Bacteria is divided by
    a. Binary fission
    b. Mitosis
    c. Budding
    d. Spore formation

2. The methods to measure bacterial growth dynamics do not involve
    a. Enumeration of cells
    b. Enumeration of colonies
    c. Initial seeding density
    d. Bulk methods such as turbidity measurement

3. The phases of bacterial growth curve in the proper sequence is
    a. Lag, log, stationary, death
    b. Lag, stationary, log, death
    c. Stationary, lag, log, death
    d. Log, stationary, lag, death

4. The phase of bacterial culture when cells prepare to divide is
    a. Lag
    b. Log
    c. Stationary
    d. exponential

5. The phase of bacterial culture when maximum cell division occurs is
    a. Lag
    b. Log
    c. Stationary
    d. Plateau

6. The phase of bacterial culture when cell division is limited by nutrient and space availability is
   a. Lag
   b. Log
   c. Stationary
   d. Plateau

7. Growth factors important for bacterial growth do not include
   a. Amino acids
   b. Vitamins
   c. Carbon source
   d. purines

8. Essential trace elements important for bacterial growth do not include
   a. Magnesium
   b. Carbon
   c. Iron
   d. Manganese

9. The optimal pH range for the growth of pathogenic bacteria is
   a. 7.2 to 7.4
   b. 6.3 to 7.0
   c. 7.6 to 8.2
   d. Exactly 7.2

10. The optimal temperature for the growth of pathogenic bacteria is
    a. 31°C
    b. 34°C
    c. 37°C
    d. 39°C

11. The three major ways in which bacteria obtain metabolic energy do not include
    a. Fermentation
    b. Respiration
    c. Photosynthesis
    d. Photolysis

12. Bacteria which are psychrophiles cannot grow at the temperature
    a. 30°C
    b. 15°C
    c. 10°C
    d. 5°C

13. The optimal temperature for the growth of mesophilic bacteria is
    a. 31°C
    b. 34°C
    c. 37°C
    d. 39°C

14. The optimal temperature range for the growth of thermophilic bacteria is
    a. 10°C to 20°C
    b. 20°C to 30°C
    c. 40°C to 50°C
    d. 50°C to 60°C

15. The optimal temperature range for the growth of psychrotrophic bacteria is
    a. 0°C to 30°C
    b. 20°C to 40°C
    c. 40°C to 50°C
    d. 50°C to 60°C

16. Halophilic bacteria require
    a. High salt concentrations
    b. Nil salt concentration
    c. Low salt concentration
    d. The salt concentration is not a decisive factor

17. Psychrophilic bacteria can survive at
    a. -10°C
    b. 40°C
    c. 60°C
    d. 80°C

18. Capnophilic bacteria require
    a. High CO_2 concentrations
    b. Low CO_2 concentrations
    c. High O_2 concentrations
    d. The CO_2 concentration is not a decisive factor

19. The quantification of infective viral particles can be performed by
    a. Inhibition assay
    b. Cytotoxic assays
    c. Plaque assay
    d. Turbidity method

20. The plaque assay to quantify the number of infective viral particles can utilize
    a. Bacteria cells
    b. Animal cells
    c. Plant cells
    d. Bacterial and animal cells

21. The target cell type that a virus infects depends on
    a. Surface proteins of the virus
    b. Surface receptors on the host cell
    c. Both (a) and (b)
    d. The HLA molecules on the host cells

22. The correct order of sequences that occur in the viral lytic cycle is
    a. Attachment, entry, replication, release
    b. Entry, attachment, release, replication
    c. Release, attachment, entry, replication
    d. Entry, release, replication, attachment

23. Budding from host cells is typical of the virus group
    a. Enveloped animal viruses
    b. Animal viruses
    c. Enveloped viruses
    d. Plant viruses

24. Lysogeny, a process of integration of viral genome into the host cell genome is typical of
    a. Animal viruses
    b. Plant viruses
    c. Bacteriophages
    d. Insect viruses

25. The prophage, as can form by the integration of viral genome into the host cell genome is typical of
    a. Animal viruses
    b. Plant viruses

    c. Bacteriophages
    d. Insect viruses

26. The viruses which have the capacity for lysogeny and the formation of prophage are appropriately termed as
    a. Animal viruses
    b. Plant viruses
    c. Bacteriophages
    d. Temperate phages

27. The time period when all the new viral progeny are formed but are not yet assembled within the host cell is
    a. Eclipse period
    b. Penetration phase
    c. Biosynthesis phase
    d. Maturation phase

28. The type of viruses which multiply in the host cell cytoplasm is
    a. ssDNA viruses
    b. dsDNA viruses
    c. All RNA viruses
    d. Only ssRNA viruses

29. Using reverse transcriptase enzyme for replication is typical of
    a. Retroviruses
    b. adenoviruses
    c. Bacteriophages
    d. Temperate phages

30. Axenic bacterial cultures refer to
    a. Pure cultures
    b. Mixed cultures
    c. A population of all bacterial types in a sample
    d. Large-scale cultures

31. Bacterial cultures are generally used for all applications except
    a. Selection of specific strains
    b. Differentiate among various stains
    c. Enrichment of select strains
    d. Obtaining co-cultures

32. Several types of bacterial species can be grown using
    a. Enriched medium
    b. Defined medium
    c. General purpose medium
    d. Indicator medium

33. The type of medium used for growing fastidious heterotrophs is
    a. Enriched medium
    b. Defined medium
    c. General purpose medium
    d. Indicator medium

34. Fastidious heterotrophic bacteria require
    a. Simple nutrition
    b. Complex nutrition
    c. Select nutrients
    d. Specific growth factors

35. A specific type of bacterial species can be grown using
    a. Enriched medium
    b. Defined medium
    c. Selective medium
    d. Indicator medium

36. Inhibition of the growth of undesired bacterial strains can be achieved using
    a. Enriched medium
    b. Defined medium
    c. Selective medium
    d. Indicator medium

37. Bacterial strain/species can be distinguished using
    a. Enriched medium
    b. Differential medium
    c. Selective medium
    d. Indicator medium

38. In a test-tube culture, the position where obligate aerobes grow is
    a. At the upper layers of the medium
    b. Mostly at the upper layer, but also distributed throughout the medium
    c. At the bottom of the medium
    d. Distributed evenly throughout the medium

39. In a test-tube culture, the position where facultative anaerobes grow is
    a. At the upper layers of the medium
    b. Mostly at the upper layer, but also distributed throughout the medium
    c. At the bottom of the medium
    d. Distributed evenly throughout the medium

40. In a test-tube culture, the position where obligate anaerobes grow is
    a. At the upper layers of the medium
    b. Mostly at the upper layer, but also distributed throughout the medium
    c. At the bottom of the medium
    d. Distributed evenly throughout the medium

41. In a test-tube culture, the position where aerotolerant anaerobes grow is
    a. At the upper layers of the medium
    b. Mostly at the upper layer, but also distributed throughout the medium
    c. At the bottom of the medium
    d. Distributed evenly throughout the medium

42. In a test-tube culture, the position where microaerophiles grow is
    a. Below the upper layer of the medium
    b. At the upper layers of the medium
    c. Mostly at the upper layer, but also distributed throughout the medium
    d. At the bottom of the medium

43. The MacConkey agar has a color indicator which shows changes in
    a. Temperature
    b. pH
    c. O_2 level
    d. CO_2 level

44. A general medium that can be used to culture most bacterial species is
    a. Yeast malt agar
    b. Sabouraud dextrose agar
    c. Tryptic soy agar
    d. Nutrient Agar

45. A general medium that can be used to culture most fastidious bacterial species is
    a. Yeast malt agar
    b. Sabouraud dextrose agar
    c. Tryptic Soy Agar

    d. Nutrient Agar

46. Examples of basal media do not include
    a. Nutrient broth
    b. Nutrient agar
    c. MacConkey agar
    d. Peptone water

47. An example of enriched medium for bacterial culture is
    a. Nutrient broth
    b. Blood agar
    c. MacConkey agar
    d. Peptone water

48. An examples of enriched medium for bacterial culture is
    a. Nutrient broth
    b. Nutrient agar
    c. MacConkey agar
    d. Lowenstein-Jensen medium

49. Examples of selective media do not include
    a. Blood agar
    b. Lowenstein-Jensen medium
    c. MacConkey agar
    d. Tellurite medium

50. Examples of indicators as can be used for bacterial culture medium do not include
    a. Blood
    b. Neutral red
    c. Peptone
    d. Tellurite

51. An example of bacterial transport medium is
    a. MacConkey agar
    b. Lowenstein-Jensen medium
    c. Tellurite medium
    d. Cary-Blair medium

52. An example of bacterial transport medium is
    a. MacConkey agar

b. Amies medium
c. Tellurite medium
d. Lowenstein-Jensen medium

53. An example of bacterial transport medium is
    a. MacConkey agar
    b. Stuart medium
    c. Tellurite medium
    d. Lowenstein-Jensen medium

54. Bacterial transport media do not include
    a. MacConkey agar
    b. Stuart medium
    c. Amies medium
    d. Cary-Blair medium

55. An example of bacterial storage medium is
    a. MacConkey agar
    b. Egg saline medium
    c. Tellurite medium
    d. Lowenstein-Jensen medium

56. An example of bacterial storage medium is
    a. Chalk cooked meat broth
    b. Stuart medium
    c. Tellurite medium
    d. Lowenstein-Jensen medium

57. The most appropriate and recommended medium for performing antimicrobial disc diffusion sensitivity is
    a. Nutrient broth
    b. Stuart medium
    c. Nutrient agar
    d. Mueller Hinton agar

58. The most appropriate and recommended medium for studying bacterial fermentation is
    a. Nutrient broth
    b. Stuart medium
    c. Hiss's serum water medium
    d. Mueller Hinton agar

59. The component in Lowenstein-Jensen medium that inhibits growth of bacteria other than mycobacteria is
    a. Egg
    b. Malachite green
    c. Glycerol
    d. Lactose

60. The component in Dubos medium that induces dispersed growth of tubercle bacilli is
    a. Tween 80
    b. Malachite green
    c. Glycerol
    d. Lactose

61. The component in Dubos medium that promotes rapid growth of tubercle bacilli is
    a. Tween 80
    b. Asparagine
    c. Bovine albumin
    d. Salts

62. The medium that is used for rapid diagnosis of Diphtheria bacilli is
    a. Eosin-methylene blue agar
    b. Loeffler serum
    c. Desoxycholate citrate agar
    d. Tellurite blood agar

63. The medium that is used for selecting *Cotynebacterium diphtheria* is
    a. Eosin-methylene blue agar
    b. Loeffler serum
    c. Desoxycholate citrate agar
    d. Tellurite blood agar

64. The medium that is used for selecting enteric Gram-negative bacilli is
    a. Eosin-methylene blue agar
    b. Loeffler serum
    c. Desoxycholate citrate agar
    d. Tellurite blood agar

65. A bacterial culture medium that is used for selecting Salmonella from stool samples is
    a. Eosin-methylene blue agar
    b. Loeffler serum
    c. Desoxycholate citrate agar
    d. Tellurite blood agar

66. A bacterial culture medium that is used for selecting Salmonella from stool samples is
    a. Eosin-methylene blue agar
    b. Loeffler serum
    c. Tetrathionate broth
    d. Tellurite blood agar

67. A bacterial culture medium that is used for selecting Vibrio species from stool samples is
    a. Eosin-methylene blue agar
    b. Thiosulphate-citrate-bile-sucrose agar
    c. Tetrathionate broth
    d. Tellurite blood agar

68. A bacterial culture medium that is used for transportation of stool samples is
    a. Eosin-methylene blue agar
    b. Thiosulphate-citrate-bile-sucrose agar
    c. Cary-Blair medium
    d. Tellurite blood agar

69. A bacterial differential slope medium that is used for identification of enteric bacteria is
    a. Kligler iron agar
    b. Thiosulphate-citrate-bile-sucrose agar
    c. Cary-Blair medium
    d. Tellurite blood agar

70. The bacterial medium that is used for culturing *Bordetella pertussis* is
    a. Kligler iron agar
    b. Thiosulphate-citrate-bile-sucrose agar
    c. Cary-Blair medium
    d. Bordet-Gengou medium

71. The major methods for culturing viruses do not include
    a. Inoculation into animal hosts
    b. Inoculation into embryonated eggs
    c. Liquid and semisolid defined media
    d. Cell cultures

72. The inoculation of embryonated eggs at the chorioallantoic membrane site is most appropriate for culturing
    a. Poxvirus
    b. Influenza virus
    c. Mumps virus
    d. Yellow fever virus

73. The inoculation of embryonated eggs at the allantoic cavity site is most appropriate for culturing
    a. Poxvirus
    b. Influenza virus, Yellow fever virus and rabies virus
    c. Smallpox virus
    d. Mumps virus

74. The inoculation of embryonated eggs at the amniotic sac site is most appropriate for culturing
    a. Poxvirus
    b. Influenza virus and mumps virus
    c. Rabies virus
    d. Yellow fever virus

75. The gold standard for culturing, purifying and identification of viruses as on today is by using
    a. Embryonated eggs
    b. Laboratory animals
    c. Specific animal organs
    d. Cell cultures

76. The first virus to be cultured using cell cultures is
    a. Vaccinia virus
    b. Yellow fever virus
    c. Small pox virus
    d. Influenza virus

77. The first virus to have been cultured and isolated using cell cultures towards vaccine preparation is
    a. Vaccinia virus
    b. Yellow fever virus
    c. Polio virus
    d. Influenza virus

78. The growth of viruses in cell cultures can be identified by observing
    a. Cytotoxicity
    b. Genotoxicity
    c. Cytopathic effect
    d. Changes in staining pattern

79. The cytopathic effects in the cell lines after inoculating with viruses typically occurs after
    a. 1-3 days
    b. 2-5 days
    c. 5-10 days
    d. 23 days

80. The cytopathic effect as observed as grape-like clusters in A549 or RhMK cells is produced by
    a. Rhinovirus
    b. Influenza virus
    c. Cytomegalovirus
    d. Adenovirus

81. The cytopathic effect as observed as rounded large cells in A549 or RhMK cells is produced by
    a. Rhinovirus
    b. Herpes simplex virus
    c. Cytomegalovirus
    d. Adenovirus

82. The cytopathic effect as observed as rounding and degeneration in A549 or RhMK cells is produced by
    a. Rhinovirus
    b. Influenza virus
    c. Cytomegalovirus
    d. Adenovirus

83. The Cytopathic effect as observed as granulations in A549 or RhMK cells is produced by
    a. Rhinovirus
    b. Influenza virus
    c. Cytomegalovirus
    d. Adenovirus

84. The most appropriate cell line to culture herpes simplex virus as can be visualized by well induced cytopathic effects is
    a. Fibroblasts
    b. A549 cells
    c. RhMK cells
    d. All of the above

85. The most appropriate cell line to culture adenovirus as can be visualized by well induced cytopathic effects is
    a. Fibroblasts
    b. A549 cells
    c. RhMK cells
    d. All of the above

86. The most appropriate cell line to culture influenza virus as can be visualized by well induced cytopathic effects is
    a. Guinea pig erythrocytes
    b. A549 cells
    c. RhMK cells
    d. All of the above

87. The most appropriate cell line to culture parainfluenza virus as can be visualized by well induced cytopathic effects is
    a. Guinea pig erythrocytes
    b. A549 cells
    c. RhMK cells
    d. All of the above

88. The most appropriate cell line to culture rhinovirus as can be visualized by well induced cytopathic effects is
    a. Fibroblasts
    b. A549 cells
    c. RhMK cells
    d. HeLa cells

89. The most appropriate cell line to culture cytomegalovirus as can be visualized by well induced cytopathic effects is
   a. Fibroblasts
   b. A549 cells
   c. RhMK cells
   d. HeLa cells

90. The most appropriate cell line to culture rubella virus as can be visualized by well induced cytopathic effects is
   a. Hep-2 cells
   b. Vero cells
   c. BHK-21 cells
   d. HeLa cells

91. Viruses such as influenza virus and mumps virus that do not induce visible cytopathic effects in cell lines can be detected by
   a. Flow-sorting
   b. Western blotting
   c. Hemadsorption
   d. Hemolysis

92. Co-cultures of 549 and MRC-5 cell lines is useful for detecting
   a. Cytomegalo virus (CMV)
   b. Herpes simplex virus (HSV)
   c. Both CMV and HSV
   d. Influenza and pox viruses

93. The transgenic cell lines that proved to be useful for the culturing and identification of human immunodeficiency virus (HIV) include
   a. CD4$^+$ lymphoid cell line
   b. HT4 cells
   c. Both CD4$^+$ lymphoid cell line and HT4 cells
   d. A549 cells

94. Bacteriophages being cultured can be identified by their capacity to induce
   a. Plaques
   b. Genotoxicity
   c. Granules
   d. Cytopathic effects

95. End-point dilution is a method to identify the viral property of
    a. Host cell attachment
    b. Host cell suitability
    c. Lethal dilution
    d. Lytic nature

CHAPTER 11

Microbial Metabolism

1. The biochemical similarity of bacterial cells with eukaryotic cells was first recognized by
   a. Kluyver and Donker
   b. Metchnikoff
   c. Louis Pasteur
   d. Davine and Holmes

2. The most common and fundamental feature of bacterial metabolism is the
   a. Transfer of oxygen ions
   b. Transfer of hydrogen ions
   c. Transfer of carbon ions
   d. Transfer of neutrons

3. The primary source of energy of bacterial cells is
   a. Adenosine di phosphate (ADP)
   b. Co-enzyme A
   c. Carbon
   d. Adenosine triphosphate (ATP)

4. The most important coenzymes for bacterial metabolism are
   a. Minerals
   b. Vitamins of B complex
   c. Vitamin C
   d. Vitamin K derivatives

5. The unity theory of biochemistry was proposed by
   a. Donker
   b. Kluyver
   c. Holmes
   d. Davine

6. The types of bacteria based on their metabolic pathways do not include
   a. Heterotrophs
   b. Autotrophs
   c. Phototrophs
   d. Chemotrophs

7. The chemoorganotrophic bacteria are
   a. Heterotrophs
   b. Autotrophs
   c. Phototrophs
   d. Chemolithotrophs

8. The chemolithotrophic bacteria are
   a. Heterotrophs
   b. Autotrophs
   c. Phototrophs
   d. Chemolithotrophs

9. The phototrophic bacteria are
   a. Heterotrophic bacteria
   b. Autotrophic bacteria
   c. Photosynthetic bacteria
   d. Chemolithotrophic bacteria

10. The heterotrophic bacteria are also known as
    a. Chemoorganotrophs
    b. Autotrophs
    c. Chemolithotrophs
    d. Phototrophs

11. The autotrophic bacteria are also known as
    a. Chemoorganotrophs
    b. Autotrophs
    c. Chemolithotrophs
    d. Phototrophs

12. The photosynthetic bacteria are also known as
    a. Chemoorganotrophs
    b. Autotrophs
    c. Chemolithotrophs
    d. Phototrophs

13. All pathogenic bacteria are
    a. Chemoorganotrophs
    b. Autotrophs
    c. Heterotrophs
    d. Phototrophs

14. The heterotrophic bacteria obtain energy by
    a. Oxidation of inorganic compounds
    b. Oxidation of organic compounds
    c. Using sunlight
    d. Carbon dioxide

15. The autotrophic bacteria obtain energy by
    a. Oxidation of inorganic compounds
    b. Oxidation of organic compounds
    c. Using sunlight
    d. Carbon dioxide

16. The hydrogen source for photoorganotrophic bacteria is
    a. H_2S
    b. H_2
    c. Organic compounds
    d. Inorganic compounds

17. The cyanobacteria obtain energy by
    a. Oxidation of inorganic compounds
    b. Oxidation of organic compounds
    c. Using sunlight
    d. Carbon dioxide

18. The carbon source for heterotrophic bacteria is
    a. H_2S
    b. CO_2
    c. Organic compounds
    d. Inorganic compounds

19. The biochemical reactions which serve for energy production while producing important precursor molecules are known as
    a. Catabolic reactions
    b. Anabolic reactions
    c. Endorganic reactions

    d. Amphibolic reactions

20. The most abundant and commonly used metabolic substrate of heterotrophs is
    a. Pyruvate
    b. Glucose
    c. Glycogen
    d. Lipids

21. Typically, the number of ATP molecules formed by the oxidation of 1 glucose molecule during aerobic cellular respiration is
    a. 24
    b. 28
    c. 34
    d. 38

22. Typically, the number of ATP molecules formed by the oxidation of 1 glucose molecule during anaerobic cellular respiration is
    a. 2
    b. 4
    c. 6
    d. 8

23. Typically, the number of calories obtained by the oxidation of 1 mole of glucose molecule during aerobic cellular respiration is
    a. 380,000
    b. 280,000
    c. 340,000
    d. 288,800

24. The efficiency of a typical cellular respiration cycle is
    a. 40%
    b. 45%
    c. 55%
    d. 65%

25. The glucose oxidation process to be completed has the steps which do not include
    a. Glycolytic pathways
    b. Glyconeogenesis pathways
    c. Krebs cycle
    d. Electron transport and oxidative phosphorylation

26. The catabolic pathway for utilizing glucose which is seen exclusively in obligate aerobic bacteria is
    a. Entner-Doudoroff pathway
    b. Heterofermentative pathway
    c. Phosphoketolase pathway
    d. Phosphofructokinase pathway

27. The acceptor of the terminal electron during fermentation is
    a. An inorganic compound
    b. An organic compound
    c. ADP
    d. ATP

28. The bacteria which utilize inorganic compounds as their only source of energy are
    a. Autotrophs
    b. Phototrophs
    c. Photoorganotrophs
    d. Heterotrophs

29. Anaerobic bacteria are
    a. Autotrophs
    b. Phototrophs
    c. Photoorganotrophs
    d. Heterotrophs

30. The electron acceptors of anaerobic (heterotrophic) bacteria do not include
    a. SO_4^{2-}
    b. O_2
    c. NO_3^-
    d. CO_2

31. The terminal electron acceptor of anaerobic respires is
    a. SO_4^{2-}
    b. O_2
    c. NO_3^-
    d. CO_2

32. The essential step in the nitrogen cycle is
    a. Ammonification
    b. Nitrification

    c. Denitrification
    d. Nitrogen fixing

33. Mixotrophic bacteria can obtain carbon from
    a. Organic compounds
    b. CO_2
    c. Both organic compounds and CO_2
    d. Inorganic compounds

34. Lithotrophic bacteria obtain reducing equivalents from
    a. Organic compounds
    b. Inorganic compounds
    c. CO_2
    d. Both organic compounds and CO_2

35. The energy sources of methylotrophic bacteria are
    a. C1 compounds
    b. C2 compounds
    c. C3 compounds
    d. C4 compounds

36. The electron acceptor in bacteria with acetogenesis type of metabolism is
    a. Fe^{3+}
    b. CO_2
    c. H_2S
    d. N_2O

37. A common terminal electron acceptor in bacteria with anaerobic metabolism is
    a. Fe^{3+}
    b. CO_2
    c. H_2S
    d. N_2O

38. The type of bacterial metabolic pathway that is consider primitive is
    a. Aerobic respiration
    b. Anaerobic respiration
    c. Methanogenesis
    d. Anoxygenic photosynthesis

39. The metabolic pathway that predominantly occurs in bacteria belonging to the Genus *Pseudomonas* is
    a. Embden-Meyerhoff-Pamas-pathway
    b. Entner-Doudoroff pathway
    c. Phosphoketolase pathway
    d. Hexose monophosphate pathway

40. The metabolic pathway that predominantly occurs in bacteria belonging to the *Bifidobacterium* group is
    a. Embden-Meyerhoff-Pamas-pathway
    b. Entner-Doudoroff pathway
    c. Phosphoketolase pathway
    d. Hexose monophosphate pathway

41. The metabolic pathway that predominantly occurs in bacteria belonging to the *Leuconostoc* group is
    a. Embden-Meyerhoff-Pamas-pathway
    b. Entner-Doudoroff pathway
    c. Phosphoketolase pathway
    d. Hexose monophosphate pathway

42. The metabolic pathway that predominantly occurs in bacteria and is referred to as the "classic" glycolytic pathway is
    a. Embden-Meyerhoff-Pamas-pathway
    b. Entner-Doudoroff pathway
    c. Phosphoketolase pathway
    d. Hexose monophosphate pathway

CHAPTER 12

Microbial Genetics

1. The chromosome in most bacteria is
    a. Circular
    b. Linier
    c. Multiple circular structures
    d. Multiple linier structures

2. The bacterial genome consists of
    a. Circular DNA
    b. RNA
    c. Both DNA and RNA
    d. Linier RNA

3. The bacterial genome is located in
    a. Nucleus
    b. Nucleoid
    c. Plasmid
    d. Nucleolus

4. The extra-chromosomal DNA in bacteria is the
    a. Transposon
    b. F-factor
    c. Plasmid
    d. Operon

5. Repression is the process by which gene expression is
    a. Suppressed when not required
    b. Enhanced when required
    c. All genes are expressed
    d. All genes are suppressed

6. Induction is the process by which gene expression is
   a. Suppressed when not required
   b. Switched on when required
   c. All genes are expressed
   d. All genes are suppressed

7. Select genes expression are suppressed when not required by
   a. Repressors
   b. Promoters
   c. Inducers
   d. operators

8. Select genes expression are expressed when required by
   a. Repressors
   b. Promoters
   c. Inducers
   d. Operators

9. The gene sequences of bacteria that can code for several proteins as can be utilized for a specific purpose is known as
   a. Operons
   b. Repressors
   c. Inducers
   d. Enhancers

10. Plasmid replication in bacteria occurs
    a. Dependent on the chromosomal replication
    b. Simultaneous with the chromosomal replication
    c. Independent of chromosomal replication
    d. Controlled by chromosomal replication

11. The short sequences of DNA in the bacterial genome that migrate between different loci are
    a. Plasmids
    b. Transposons
    c. Operons
    d. Promoters

12. Genetic transfer in bacteria occurs by processes which do not include
    a. Conjugation
    b. Transduction
    c. Transformation
    d. Transfection

13. The major process by which plasmids are transferred in bacteria is
    a. Conjugation
    b. Transduction
    c. Transfection
    d. Transformation

14. The F factor is important for the process
    a. Conjugation
    b. Transduction
    c. Transfection
    d. Transformation

15. The phage-mediated genetic material transfer process in bacteria is
    a. Conjugation
    b. Transduction
    c. Transfection
    d. Transformation

16. The process by which naked DNA is directly taken up by a bacterial cell from the surroundings is
    a. Conjugation
    b. Transduction
    c. Transfection
    d. Transformation

17. Essential self-replicating genetic material of bacteria are known as
    a. Transposons
    b. Replicons
    c. Plasmids
    d. F-factor

18. Non-essential replicons of bacteria include
    a. Plasmids
    b. Circular chromosomal DNA
    c. The bacterial chromosome

d. Replicating genetic element

19. Most bacterial plasmids are
    a. Super coiled circular structures
    b. Linier single-stranded structures
    c. Essential gene-containing structures
    d. Structures that contain genes for cell viability

20. Bacteriophage genome consists of
    a. DNA
    b. RNA
    c. DNA or RNA
    d. Host DNA

21. Of the nineteen families of bacteriophages currently recognized, the number of families with RNA as the genetic material is
    a. 2
    b. 3
    c. 4
    d. 5

22. The genome of cystoviruses that infect Gram-negative bacteria comprises of
    a. dsDNA
    b. ssDNA
    c. dsRNA
    d. ssRNA

23. The family of bacteriophages that contain linier dsDNA includes
    a. *Corticoviridae*
    b. *Plasmaviridae*
    c. *Myoviridae*
    d. *Inoviridae*

24. The family of bacteriophages that contain linier dsDNA includes
    a. *Microviridae*
    b. *Siphoviridae*
    c. *Corticoviridae*
    d. *Plasmaviridae*

25. The family of bacteriophages that contain linier dsDNA includes
    a. *Podoviridae*

  b. *Leviviridae*
  c. *Corticoviridae*
  d. *Plasmaviridae*

26. The family of bacteriophages that contain linier dsDNA includes
  a. *Corticoviridae*
  b. *Microviridae*
  c. *Cystoviridae*
  d. *Tectiviridae*

27. The family of bacteriophages that contain circular dsDNA includes
  a. *Myoviridae*
  b. *Siphoviridae*
  c. *Corticoviridae*
  d. *Cystoviridae*

28. The family of bacteriophages that contain circular dsDNA includes
  a. *Plasmaviridae*
  b. *Myoviridae*
  c. *Siphoviridae*
  d. *Cystoviridae*

29. The family of bacteriophages that contain circular ssDNA includes
  a. *Myoviridae*
  b. *Tectiviridae*
  c. *Siphoviridae*
  d. *Microviridae*

30. The family of bacteriophages that contain circular ssDNA includes
  a. *Inoviridae*
  b. *Siphoviridae*
  c. *Corticoviridae*
  d. *Cystoviridae*

31. The family of bacteriophages that contain segmented dsRNA includes
  a. *Myoviridae*
  b. *Siphoviridae*
  c. *Corticoviridae*
  d. *Cystoviridae*

32. The family of bacteriophages that contain linier ssRNA includes
    a. *Leviviridae*
    b. *Inoviridae*
    c. *Corticoviridae*
    d. *Cystoviridae*

CHAPTER 13

Microbial Control

1. A treatment process that renders a material free of any viable microbe is known as
   a. Sterilization
   b. Disinfection
   c. Inhibition
   d. Fumigation

2. A treatment process that removes or kills all microbes from a substance is most appropriately known as
   a. Sterilization
   b. Disinfection
   c. Inhibition
   d. Fumigation

3. A treatment process that renders an inanimate object free of living microbes is appropriately known as
   a. Sterilization
   b. Disinfection
   c. Inhibition
   d. Fumigation

4. A treatment process that renders living substances (such as skin) free of living microbes is appropriately known as
   a. Sterilization
   b. Disinfection
   c. Antisepsis
   d. Swabbing

5. A treatment process that removes microbes and their potential harmful metabolic products is appropriately known as
   a. Decontamination
   b. Disinfection
   c. Antisepsis
   d. Sterilization

6. A treatment process that is practiced to remove microbes from living substances (such as skin) is appropriately known as
   a. Sterilization
   b. Decontamination
   c. Degerming
   d. Swabbing

7. A treatment process that is practiced to minimize the number of microbes from inanimate material is appropriately known as
   a. Disinfection
   b. Decontamination
   c. Sanitization
   d. Swabbing

8. Procedures that are followed with meticulous care to prevent microbial contamination are known as
   a. Aseptic methods
   b. Clean methods
   c. Standard operating procedures
   d. Good laboratory practices

9. Microbial inhibition leads to
   a. Killing of all microbes present
   b. Killing of 50% of the microbial population
   c. Halting the microbial population growth
   d. Halting of 50% of the microbial population growth

10. The effects of most disinfectants and sterilants is in
    a. Acting on all microbes present in a non-specific manner
    b. Killing of 50% of the microbial population
    c. Halting the microbial population growth
    d. Halting of 50% of the microbial population growth

11. The temperature required to kill 90% of microbes in a specific time period is known as
    a. Decimal reduction time (D)-value
    b. Phenol coefficient (PC)-value
    c. Z-value
    d. Thermal death point (TDP)-value

12. The temperature required to kill 100% of microbes from a 24-hour broth culture period is known as
    a. Decimal reduction time (D)-value
    b. Phenol coefficient (PC)-value
    c. Z-value
    d. Thermal death point (TDP)-value

13. The temperature required to kill 90% of microbes is known as
    a. Decimal reduction time (D)-value
    b. Phenol coefficient (PC)-value
    c. Z-value
    d. Thermal death point (TDP)-value

14. The evaluation of effectiveness of disinfectants is usually measured as
    a. Decimal reduction time (D)-value
    b. Phenol coefficient (PC)-value
    c. Z-value
    d. Thermal death point (TDP)-value

15. An example of a pressured steam sterilizer is
    a. Autoclave
    b. Incubator
    c. Laminar air flow
    d. Hot air oven

16. The ideal pressure for effective sterilization using an autoclave is
    a. 5 psi
    b. 10 psi
    c. 15 psi
    d. 20 psi

17. The ideal temperature for effective sterilization using an autoclave is
    a. 100^0C
    b. 110^0C
    c. 121^0C
    d. 140^0C

18. Pasteurization is a disinfection process that uses
    a. Chemicals
    b. Heat
    c. Flame

  d. Moisture

19. The ideal temperature for effective sterilization using dry heat is
  a. 110⁰C
  b. 121⁰C
  c. 140⁰C
  d. 180⁰C

20. The ideal time period required for effective sterilization using dry heat at 180⁰C is
  a. 1 hour
  b. 2 hours
  c. 3 hours
  d. 4 hours

21. Cold temperature is usually
  a. Bacteriostatic
  b. Bacteriocidal
  c. Fungicidal
  d. Germicidal

22. Industrial sterilization by ionizing radiation uses
  a. A gamma source
  b. Cesium
  c. Alpha radiation
  d. Beta radiation

23. Sunlight has anti-microbial effects through
  a. Photo-oxidation
  b. Reduction
  c. Alkylation
  d. Conversion

24. Ultrasound has anti-microbial effects through
  a. Cavitation
  b. Oxidation
  c. Reduction
  d. Alkylation

25. An important component of safety hoods that effectively filters microbes is
  a. HEPA filter

    b. Nylon filter
    c. UV lamp
    d. Bunsen burner

26. Preventing infections while working with infective strains of microbes is achieved by using
    a. Negative pressure airflow systems
    b. Positive pressure airflow systems
    c. Unidirectional outward flow systems
    d. Gloves

27. Maintaining sterile conditions and prevention of contaminations while working with microbes is achieved by using
    a. Negative pressure airflow systems
    b. Positive pressure airflow systems
    c. Unidirectional outward flow systems
    d. Gloves

28. Halogens that can be used to control the spread of microbes do not include
    a. Chlorine
    b. Iodine
    c. Sodium
    d. Bromine

29. The major component of tincture is
    a. Chlorine
    b. Iodine
    c. Sodium
    d. Bromine

30. The most common compound used for fumigating work areas that involve microbial cultures is
    a. Formaldehyde
    b. Iodine
    c. Bromine
    d. Chloroform

31. The most common compound used along with formaldehyde for fumigating work areas that involve microbial cultures is
    a. Chlorine
    b. Iodine

   c. Bromine
   d. Potassium permanganate

32. The compound that is bound to a surfactant carrier for the control of microbes is
    a. Chlorine
    b. Iodine
    c. Iodophore
    d. Bromine

33. Iodophores are normally
    a. Water insoluble
    b. Water soluble
    c. Soluble in alcohols
    d. Soluble in oils

34. One most commonly used alcohol as a disinfectant is
    a. Propanol
    b. Butanol
    c. Isopropanol
    d. Methanol

35. One most commonly used alcohol as a disinfectant is
    a. Propanol
    b. Butanol
    c. Isopropanol
    d. Ethanol

36. The % of formaldehyde in formalin is
    a. 23%
    b. 37%
    c. 46%
    d. 73%

37. Heat as an antimicrobial agent best works at
    a. Neutral pH
    b. High pH
    c. Low pH
    d. pH has no influence

38. The usual amount of time required to kill all vegetative bacteria and viruses by boiling is
    a. 5 minutes
    b. 10 minutes
    c. 15 minutes
    d. 20 minutes

39. The temperature that is used for the process of classic pasteurization is
    a. 36^0C
    b. 46^0C
    c. 63^0C
    d. 76^0C

40. The time duration required for the process of classic pasteurization is
    a. 15 minutes
    b. 30 minutes
    c. 45 minutes
    d. 60 minutes

41. The temperature that is used for the process of HTST pasteurization is
    a. 72^0C
    b. 63^0C
    c. 43^0C
    d. 34^0C

42. The time duration required for the process of HTST pasteurization is
    a. 15 minutes
    b. 15 seconds
    c. 30 minutes
    d. 30 seconds

43. Dry heat kills microbes by
    a. Alkylation
    b. Reduction
    c. Hydrogenation
    d. Oxidation

44. The pore size of most HEPA filters is
    a. 0.1 um
    b. 0.3 um
    c. 0.4 um

d. 0.6 um

45. The pore size that can filter viruses is
    a. 0.01 um
    b. 0.1 um
    c. 0.03 um
    d. 0.3 um

46. Moist heat functions for microbial control by
    a. Oxidation
    b. DNA denaturation
    c. Interfering with microbial metabolism
    d. Protein denaturation

47. Autoclaving functions for microbial control by
    a. Protein denaturation
    b. Oxidation
    c. DNA denaturation
    d. Interfering with microbial metabolism

48. Pasteurization functions for microbial control by
    a. Protein denaturation
    b. Oxidation
    c. DNA denaturation
    d. Interfering with microbial metabolism

49. Protein denaturation as can be used for microbial control can be achieved by
    a. Boiling
    b. Direct flaming
    c. Cold temperature
    d. Ionizing radiation

50. Protein denaturation as can be used for microbial control can be achieved by
    a. Deep-freezing
    b. Autoclaving
    c. Cold temperature
    d. Ionizing radiation

51. Protein denaturation as can be used for microbial control can be achieved by
    a. Deep-freezing
    b. High pressure

  c. Pasteurization
  d. Ionizing radiation

52. Destruction of metabolism as can be used for microbial control can be achieved by
  a. Boiling
  b. Desiccation
  c. Cold temperature
  d. Ionizing radiation

53. Inducing DNA damages as can be used for microbial control can be achieved by
  a. Boiling
  b. Desiccation
  c. Cold temperature
  d. Radiation

54. Inducing plasmolysis as can be used for microbial control can be achieved by
  a. Osmotic pressure
  b. Desiccation
  c. Cold temperature
  d. Lyophilization

55. Phenolics as disinfectants are rendered more effective by the addition of
  a. Chlorine
  b. Enzymes
  c. Detergents
  d. Alcohol

56. The phenolics disinfectants target
  a. DNA
  b. Lipids
  c. Spindle fibers
  d. Plasma membrane

57. The mechanism of action of phenolics disinfectants is by
  a. Enzyme inactivation
  b. DNA damage
  c. Prevention of cell division
  d. Disturbing protein synthesis

58. The germicide that functions by generating hypochlorous acid in water is
    a. Alcohol
    b. Chlorine
    c. Iodine
    d. Hydrogen peroxide

59. The germicide that functions by solubilizing lipids and denaturing proteins is
    a. Alcohol
    b. Chlorine
    c. Iodine
    d. Hydrogen peroxide

60. The germicides that function by inducing an oligodynamic effects are
    a. Alcohols
    b. Chlorine
    c. Phenolics
    d. Heavy metals

61. The % of silver nitrate to be an effective disinfectant is
    a. 1%
    b. 2%
    c. 3%
    d. 4%

62. The heavy metal of choice that is included in mouthwashes is
    a. Silver
    b. Zinc
    c. Selenium
    d. Copper

63. The % of glutaraldehyde that is used commonly as a chemical disinfectant is
    a. 1%
    b. 2%
    c. 3%
    d. 4%

64. An example of an effective liquid sporicidal agent is
    a. Hydrogen peroxide
    b. Benzoyl peroxide
    c. Peracetic acid
    d. Formaldehyde

CHAPTER 14

Microbial Diseases

1. Which of the following indicates the population of microorganisms that inhabit a healthy human body more or less on a permanent basis?
   a. Transient flora
   b. Probiotics
   c. Microflora
   d. Resident flora

2. An individual who has recovered from an infectious disease but continues to harbor a large number of pathogens is called
   a. Acute carrier
   b. Incubatory carrier
   c. Healthy carrier
   d. Convalescent carrier

3. In 1863, who proposed the role of Microorganisms in the origin of infections?
   a. Louis Pasteur
   b. Robert Koch
   c. Oliver Wendel Holmes
   d. Paul Ehrlich

4. The toxin that is heat stable and is produced by Gram-negative pathogens is:
   a. Endotoxin
   b. Exotoxin
   c. Leucocidins
   d. All of the above

5. When a disease occurs at moderately regular intervals in a steady, low-level frequency, it is a/an:
   a. Epidemic
   b. Endemic
   c. Sporadic
   d. Pandemic

6. Which among the following is NOT the specimen of choice for diagnosis of Upper Respiratory Tract (URT) infections?
   a. Pernasal swab
   b. Sinus washings
   c. Nasopharyngeal swab
   d. Bronchial washings

7. Which among the following is an opportunistic pathogen residing in the gut of a normal human that may cause enterococcal infections?
   a. *Klebsiella species*
   b. *Anaerobic bacteroides*
   c. *Escherichia coli*
   d. *Proteus mirabilis*

8. The most prevalent gram-positive bacterium in the glands of the skin is:
   a. *Propionibacterium acnes*
   b. *Moraxella species*
   c. *Corynebacterium species*
   d. *Haemophilus species*

9. Which bacterium is a non-sporing gram-positive rod?
   a. *Bacillus*
   b. *Clostridium*
   c. *Listeria*
   d. *Pseudomonas*

10. The bacterium that are pleomorphic and devoid of cell walls showing a characteristic "fried egg" appearance when cultured is the:
    a. *Mycobacterium*
    b. *Mycoplasma*
    c. *Acetobacterium*
    d. *Leuconostoc*

11. Which of the following bacteria is a gram-negative, facultative anaerobic rod?
    a. *Shigella*
    b. *Neisseria*
    c. *Legionella*
    d. *Flavobacterium*

12. The morphology of which of the following bacterium is "grape-like"

clusters?
a. *Streptococcus*
b. *Pseudomonas*
c. *Staphylococcus*
d. *Mycobacterium*

13. One of the following is an opportunistic pathogen that can cause serious problems when introduced into surgical wounds?
a. *Bacillus*
b. *Clostridium*
c. *Mycoplasmas*
d. *Staphylococcus*

14. Which among the following is a cell wall-less pleomorphic organism?
a. *Propionibacterium*
b. *Streptomyces*
c. *Mycoplasmas*
d. *Listeria*

15. Which of the following is a motile, rod-shaped bacterium that can withstand the acidic conditions of human stomach and cause peptic ulcers?
a. *Helicobacter pylori*
b. *Campylobacter jejuni*
c. *Enterococcus faecalis*
d. *Listeria monocytogens*

16. One way to distinguish species of Streptococcus from each other is on the basis of which of their following reactions?
a. Catalase test
b. Oxidase test
c. Urease test
d. Haemolysis on blood agar

17. Which of the following genera mentioned below does NOT possess species that are human pathogens?
a. *Streptococcus*
b. *Escherichia*
c. *Streptococcus*
d. *Streptomyces*

18. The bacteria responsible for Lyme disease is:

a. *Borrelia burgdorferi*
b. *Listeria monocytogens*
c. *Critispira species*
d. *Chlamydia psittaci*

19. Which pathogen is famous for its "Snapping" cell division?
    a. *Mycobacterium tuberculosis*
    b. *Streptococcus mutans*
    c. *Staphylococcus species*
    d. *Corynebacterium species*

20. A species of Clostridium that causes diarrhoea associated with antibiotic use with high communicability among compromised patients is:
    a. *C. botulinum*
    b. *C. difficile*
    c. *C. perfringens*
    d. *C. tetani*

21. Which of the following causes Bacillary dysentery?
    a. *Escherichia coli*
    b. *Shigella dysenteriae*
    c. *Enterobacter*
    d. *Salmonella species*

22. Which of the following is a STD that can cause trachoma, non-gonococcal urethritis and lymphogranuloma venereum?
    a. *Chlamydia trachomatis*
    b. *Treponema pallidum*
    c. *Neisseria gonorrhoeae*
    d. *Mycoplasma genitalium*

23. "Vampire of the Bacterial World", most unusual organism, predators of other gram-negative bacteria is which one of the following?
    a. Serratia species
    b. Pseudomona species
    c. Legionella species
    d. *Bdellovibrio species*

24. The pathogen that causes Q fever in mammals is:
    a. *Haemophilus influenzae*
    b. *Myxococcus species*

c. *Coxiella burnetii*

d. *Spiroplasmas*

25. Which of the following bacteria is rod-shaped with Acid-fast cell wall?

a. *Corynebacterium*

b. *Mycobacterium*

c. *Streptococcus*

d. *Propionibacterium*

26. One of the below mentioned bacteria is famous for its Toxic Shock Syndrome. Which one is it?

a. *Clostridium*

b. *Chlamydia*

c. *Salmonella*

d. *Staphylococcus*

27. Find the odd bacteria out from the following

a. *Proteus*

b. *Rhizobium*

c. *Salmonella*

d. *Escherichia*

28. Which among the bacteria given below causes Rocky Mountain spotted fever wherein ticks are the vectors?

a. *Rickettsia rickettsii*

b. *Coxiella burnetii*

c. *Campylobacter species*

d. *Chlamydia psittaci*

29. Which of the following bacteria is comma-shaped?

a. *Pseudomonas aeruginosa*

b. *Shigella dysenteriae*

c. *Vibrio cholerae*

d. *Yersinia pestis*

30. A microaerophilic rod (Gram negative) that is most commonly found in the gastric mucosa of humans is:

a. *Helicobacter*

b. *Peptostreptococcus*

c. *Erwinia*

d. *Citrobacter*

31. One of the following bacteria cannot reproduce on their own outside the human cells.
    a. *Wolbachia*
    b. *Corynebacterium*
    c. *Neisseria*
    d. *Chlamydia*

32. Which one of the following causes zoonoses?
    a. *Eikenella*
    b. *Neisseria*
    c. *Campylobacter*
    d. *Propionobacterium*

33. Fried-egg appearance is a characteristic of which of the following bacteria?
    a. *Vibrio*
    b. *Mycoplasma*
    c. *Legionella*
    d. *Bordetella*

34. Which bacteria is a causative agent of 'Whooping cough'?
    a. *Bordetella pertussis*
    b. *Klebsiella pneumoniae*
    c. *Haemophilus influenzae*
    d. *Mycoplasma pneumoniae*

35. Botulism is a
    a. Water-borne disease
    b. Soil-borne disease
    c. Food-borne disease
    d. Air-borne disease

36. Which bacteria causes Hansen's disease?
    a. *Mycobacterium tuberculosis*
    b. *Staphylococcus epidermidis*
    c. *Pseudomonas aeuroginosa*
    d. *Mycobacterium leprae*

37. Which bacterial species produces 'Lock-jaw' in humans?
    a. *Clostridium botulinum*
    b. *Clostridium tetani*
    c. *Salmonella typhi*

d. *Clostridium histolyticum*

38. From the following, identify the Gram-positive, club-shaped bacterium that has many metachromatic granules in its cytoplasm.
    a. *Corynebacterium diphtheriae*
    b. *Streptococcus pneumoniae*
    c. *Mycobacterium tuberculosis*
    d. *Mycobacterium bovis*

39. Which of the following organism does not cause bacterial meningitis?
    a. *Streptococcus pneumoniae*
    b. *Nocardia asteroids*
    c. *Mycobacterium tuberculosis*
    d. *Salmonella typhi*

40. Which gram positive bacterium inhibits the release of the neurotransmitter Acetylcholine, thereby causing a Neuro-muscular junctional disorder?
    a. *Clostridium botulinum*
    b. *Staphylococcus hominis*
    c. *Pseudomonas aeruginosa*
    d. *Bifidobacterium animalis*

41. Which of the following bacteria is NOT a gram positive rod?
    a. *Bacillus*
    b. *Listeria*
    c. *Clostridium*
    d. *Enterococcus*

42. One of the following species belonging to Staphylococci genera forms golden-yellow colonies. Which one is it?
    a. *S. epidermidis*
    b. *S. aureus*
    c. *S. haemolyticus*
    d. *S. pasteuri*

43. A toxin produced by certain staphylococcal strains, the *exfoliative toxin*, is manifested by:
    a. Toxic Shock Syndrome
    b. Scalded Skin Syndrome
    c. Erysipelas
    d. Impetigo

44. Infections established by which organism have a characteristic blue-green pus caused by the pigment pyocyanin?
   a. *Pseudomonas aeruginosa*
   b. *Propionibacterium acnes*
   c. *Streptococcus pyogenes*
   d. *Staphylococcus epidermidis*

45. The 'Opisthotons' spasms caused by the toxin of which bacteria can actually result in a fractured spine?
   a. *Clostridium botulinum*
   b. *Clostridium perfringenes*
   c. *Clostridium tetani*
   d. *Clostridium difficile*

46. Which toxin produced by *Clostridium botulinum* is most virulent?
   a. Type A toxin
   b. Type B toxin
   c. Type E toxin
   d. None of the above

47. Which organism is the causative agent of Acute Bacterial Endocarditis?
   a. *Streptococcus pyogenes*
   b. *Streptococcus agalactiae*
   c. *Staphylococcus epidermidis*
   d. *Staphylococcus aureus*

48. Plague or the 'Black Death' is caused by which of the following bacterium?
   a. *Yersinia pestis*
   b. *Yersinia enterocolitica*
   c. *Yersinia pseudotuberculosis*
   d. None of the above

49. Which species of Salmonella causes characteristic "Rose spots" on the skin during the second week of fever?
   a. *S. typhi*
   b. *S. enteritidis*
   c. *S. pullorum*
   d. *S. gallinarum*

50. The causative agent of Venereal syphilis in humans is:
   a. *T. carateum*

    b. *T. pertenue*
    c. *T. endemicum*
    d. *T. pallidum*

51. Which bacterium is considered to have a characteristic "fishy" putrefaciens odour?
    a. *Plesiomonas shigelloides*
    b. *Proteus mirabilis*
    c. *Erwinia herbicola*
    d. *Shigella sonnei*

52. Which of the following is NOT a transport medium for *V. cholerae*?
    a. Cary-Blair medium
    b. Venkataraman-Ramakrishnan (VR) medium
    c. Sea water (autoclaved)
    d. Stuart's medium

53. Which bacterium causes 'Red Leg Disease'?
    a. *Aeromonas hydrophila*
    b. *Plesiomonas shigelloids*
    c. *Staphylococcus aureus*
    d. *Vibrio vulnificus*

54. Brucellosis is an infection of:
    a. Cardiovascular system
    b. Nervous system
    c. Reticuloendothelial system
    d. Lymphatic system

55. The most common cases of cystitis is caused by:
    a. *Mycobacterium hominis*
    b. *Escherichia coli*
    c. *Chlamydia trachomatis*
    d. *Gardnerella vaginalis*

56. Complete haemolysis around the colonies of Streptococcal species is called:
    a. Alpha haemolysis
    b. Beta haemolysis
    c. Gamma haemolysis
    d. All of the above

57. Viruses can be viewed under which of the following microscope?
    a. Electron microscope
    b. Bright-field microscope
    c. Fluorescent microscope
    d. Polarizing microscope

58. Hepatitis A virus belongs to which family?
    a. Togaviridae
    b. Coronaviridae
    c. Arteriviridae
    d. Picornaviridae

59. To which genera does the Rabies virus (Rhabdoviridae family) belong?
    a. Vesiculovirus
    b. Lyssavirus
    c. Ephemerovirus
    d. Novirhabdovirus

60. The "Yellow Fever Virus" belongs to which group of viruses?
    a. Lagovirus
    b. Pestivirus
    c. Flavivirus
    d. Aphthovirus

61. Viruses causing influenza types A, B and C, belong to which family?
    a. Orthomyxoviridae
    b. Rhabdoviridae
    c. Arenaviridae
    d. Retroviridae

62. The causative agent of Measles
    a. Morbillivirus
    b. Rubulavirus
    c. Vesiculovirus
    d. Bunyavirus

63. Which of the following are the largest and most complex of all viruses?
    a. Poxvirus
    b. Gyrovirus
    c. Hantavirus
    d. Circovirus

64. Which of the following virus is an excellent example of 'creepers'?
    a. Poxvirus
    b. Herpes Simplex Virus
    c. EBV
    d. Hantavirus

65. The type of Nucleic acid in Rhinovirus is:
    a. Single stranded RNA
    b. Double stranded RNA
    c. Single stranded DNA
    d. Double stranded DNA

66. During which phase does replication of viral genetic material occur?
    a. Penetration
    b. Maturation
    c. Biosynthesis
    d. Release

67. One of the following is a synonym to the 'outer coat' of a virus.
    a. Spike
    b. Capsid
    c. Capsomere
    d. Capsule

68. Which one of the below mentioned viruses is responsible for Varicella (Chicken pox) in children?
    a. HSV
    b. EBV
    c. CMV
    d. VZV

69. Which of the following virus is characterized by "Slapped Cheek" (Erythema infectiosum)?
    a. Parvovirus
    b. Hepatitis B Virus
    c. Enterovirus
    d. Poliovirus

70. One of the following virus is the major causative agent of Common cold. Which is it?
    a. Rotavirus

b. Retrovirus
c. Rhinovirus
d. Reovirus

71. The genome of which of the following virus has two copies of ssRNA with Reverse Transcriptase enzyme?
a. HIV
b. HPV
c. HBV
d. HDV

72. The Family to which Lentiviruses belong is
a. Pneumovirinae
b. Arenaviridae
c. Retroviridae
d. Paramyxovirinae

73. Poliovirus belongs to which of the following family?
a. Togaviridae
b. Bornaviridae
c. Filoviridae
d. Picornaviridae

74. The family Filoviridae contains which one of the following viruses?
a. Aichi virus
b. Rubella virus
c. Sapporo virus
d. Ebola Virus

75. Select the odd one out that does NOT establish a disease in man
a. HIV
b. HBV
c. HHV
d. HDV

76. Landsteiner and Popper established the viral nature of which disease?
a. Acquired Immuno Deficiency Syndrome
b. Poliomyelitis
c. Hepatitis
d. Small Pox

77. Which virus has the presence of a distinctive triple-layered icosahedral protein capsid?
    a. Hepatitis B Virus (HBV)
    b. Human Immunodeficiency Virus (HIV)
    c. Human Papilloma Virus (HPV)
    d. Human Rota Virus (HRV)

78. Who established the presence of Rabies Virus in the brain of infected animals?
    a. Robert Koch
    b. Louis Pasteur
    c. Paul Ehrlich
    d. Edward Jenner

79. Which disease, caused by RNA virus, is characterized by non-suppurative enlargement of parotid glands?
    a. Measles
    b. Mumps
    c. Rubella
    d. Influenza

80. The 'Star-shaped' virus that causes outbreaks of diarrhea is:
    a. Norwalk Virus
    b. Sapporo Virus
    c. Adenovirus
    d. Astrovirus

81. Which mosquito-borne infection is also called "Break-bone fever"?
    a. Dengue
    b. Chikungunya
    c. Japanese Encephalitis
    d. None of the above

82. Which virus, the natural pathogens of humans and mice, causes haemorrhagic fever with Renal syndrome (HFRS)?
    a. Arbovirus
    b. Flavivirus
    c. Hantavirus
    d. Rotavirus

83. Which of the following virus is also called the "Orphan's Virus"?

    a. ECHO
    b. Rhino Virus
    c. Coxsackie Virus
    d. Polio Virus

84. One of the characteristic symptoms 'Hydrophobia' is caused by which virus?
    a. Rubella Virus
    b. Rabies Virus
    c. Rota Virus
    d. Rhino Virus

85. Which virus causes "German Measles"?
    a. Parvovirus
    b. Mumps Virus
    c. Measles Virus
    d. Rubella Virus

86. Which of the following is NOT an RNA Virus?
    a. Measles virus
    b. Mumps virus
    c. Rubella Virus
    d. Human Parvo virus

87. The 'Swine Flu' is caused by:
    a. H5N1
    b. H1N1
    c. H1N2
    d. H2N2

88. The T4 bacteriophage which is a classic example of virulent dsDNA belongs to which family?
    a. Myoviridae
    b. Retroviridae
    c. Picornaviridae
    d. Togaviridae

89. The herpes simplex virus types 1 and 2 belongs to which of the following subfamily?
    a. Alpha herpes virus
    b. Beta herpes virus
    c. Gamma herpes virus

d. None of the above

90. Which type of viruses use an enzyme called "RNA-dependent RNA polymerase" to synthesize mRNA for replicating their genomes?
    a. ssRNA
    b. dsRNA
    c. Both types of RNA viruses
    d. +strand RNA genomes

91. Which of the following characterize Negri bodies?
    a. Intracytoplasmic inclusion bodies of HSV
    b. Intranuclear inclusion bodies of HSV
    c. Intracytoplasmic inclusion bodies of Rabies Virus
    d. Intracytoplasmic inclusion bodies of HBV

92. One of the following viruses aid in forming Syncytia. Which is it?
    a. Human Immunodeficiency Virus
    b. Herpes Simplex Virus
    c. Hepatitis Virus
    d. Ebola Virus

93. Transmission of HIV is by:
    a. Blood transfusion
    b. Sexual Contact
    c. Congenital
    d. All of the above

94. Subacute Sclerosing Panencephalitis (SSPE) is a neurological complication of which virus?
    a. Measles Virus
    b. Mumps Virus
    c. Both a) and b)
    d. None of the above

95. The Edmonston-Zagreb strain in live-attenuated vaccine is used for the prophylaxis of which virus?
    a. Rubella Virus
    b. Varicella Zoster Virus
    c. Measles Virus
    d. Small Pox virus

96. Which virus, transmitted by Culex mosquitoes, causes Dengue-like illness in humans?
    a. Chikungunya virus
    b. Puumala virus
    c. West Nile virus
    d. Hantaan virus

97. Severe Acute Respiratory Syndrome (SARS) is a severe atypical pneumonia, caused by:
    a. Corona Virus
    b. Influenza Virus
    c. Respiratory Synctial Virus
    d. Pox virus

98. The tick-borne Epidemic Haemorrhagic Fever (EHF) is caused by:
    a. Hantaan virus
    b. Seoul virus
    c. Puumala virus
    d. All of the above

99. Rota virus belongs to which of the following family?
    a. Birnaviride
    b. Reoviridae
    c. Polyomaviridae
    d. Iridoviridae

100. The study of diseases established by fungi is:
    a. Medical Mycology
    b. Medical Phycology
    c. Medical Parasitology
    d. None of the above

ANNEXURE

ANSWERS TO MCQS

Part 1: Cell and Tissue Culture

Chapter 1. Types of Animal Cell Cultures and Major Discoveries

1.a 2.a 3.c 4.d 5.b 6.c 7.b 8.c 9.a 10.a 11.c 12.b 13.c 14.a 15.c 16.c 17.c 18.b 19.a 20.c 21.c 22.d 23.b 24.a 25.c 26.a 27.a 28.c 29.a 30.d 31.b 32.d 33.c 34.a 35.b 36.a 37.c 38.b 39.a 40.a 41.c 42.b 43.d 44.a 45.b 46.b 47.d 48.a 49.c 50.b 51.d 52.a 53.d 54.a

Chapter 2. Animal Cell Culture Media and Supplements

1.d 2.b 3.a 4.c 5.d 6.b 7.a 8.c 9.a 10.a 11.b 12.a 13.c 14.b 15.c 16.c 17.b 18.a 19.a 20.a 21.b 22.a 23.a 24.c 25.b 26.b 27.c 28.b 29.c 30.b 31.c 32.b 33.b 34.b 35.b 36.a 37.c 38.a 39.a 40.b 41.b 42.d 43.a 44.a 45.b 46.a 47.a 48.b 49.c 50.b 51.c 52.d 53.c 54.c 55.b 56.a 57.b 58.c 59.a 60.a 61.d 62.c 63.d 64.a 65.b 66.d 67.c 68.a 69.c 70.b 71.a 72.b 73.c 74.b 75.c 76.a 77.d 78.a 79.a 80.c 81.b 82.c 83.c 84.d 85.b 86.a 87.a 88.d 89.c 90.d

Chapter 3. Animal and Human Cell Lines

Section 3.1. Animal Cell Lines

1.a 2.c 3.a 4.c 5.d 6.c 7.b 8.a 9.b 10.c 11.b 12.a 13.a 14.c 15.a 16.c 17.b 18.a 19.b 20.d 21.b 22.c 23.d 24.d 25.a 26.a 27.c 28.b 29.b 30.a 31.d 32.b 33.c 34.c 35.a 36.a 37.b 38.b 39.c 40.a 41.b 42.a 43.c 44.a 45.a 46.b 47.d 48.b 49.c 50.b 51.c 52.b 53.a 54.b 55.a 56.b

Section 3.2. Human Cell Lines

57.a 58.c 59.b 60.a 61.d 62.b 63.d 64.b 65.c 66.a 67.c 68.c 69.a 70.b 71.b 72.a 73.a 74.a 75.d 76.b 77.b 78.a 79.a 80.c 81.b

Chapter 4. Hybridoma Technology

1.b 2.a 3.a 4.d 5.c 6.d 7.c 8.b 9.b 10.a 11.c 12.a 13.d 14.b 15.c 16.b 17.a 18.d 19.b 20.b 21.c 22.a 23.b 24.c 25.d 26.a 27.a 28.a 29.c

Chapter 5. 3D Cell Cultures

1.b 2.a 3.a 4.c 5.c 6.a 7.d 8.c 9.b 10.c 11.a 12.a 13.a 14.a 15.d 16.c 17.c 18.c 19.c 20.b 21.b 22.a 23.c 24.c 25.d 26.c 27.c 28.d 29.a 30.a 31.b 32.c 33.b 34.a 35.c 36.d 37.d 38.b 39.c 40.d 41.b

Chapter 6. Stem Cells

1.b 2.c 3.a 4.a 5.d 6.c 7.c 8.b 9.a 10.c 11.c 12.d 13.a 14.c 15.a 16.c 17.a 18.b 19.c 20.a 21.b 22.a 23.c 24.b 25.a 26.b 27.b 28.c 29.a

Part 2: Microbiology

Chapter 7. History of Microbiology

1.a 2.c 3.c 4.b 5.c 6.a 7.a 8.c 9.d 10.b 11.b 12.a 13.b 14.a 15.c 16.d 17.a 18.c 19.b 20.b 21.c 22.a 23.d 24.a 25.a 26.a 27.c 28.b 29.a 30.c 31.a 32.a 33.c 34.b 35.a 36.b 37.c 38.a 39.c 40.d 41.b 42.c 43.c 44.a 45.b 46.d 47.c 48.a 49.b 50.c 51.c 52.a 53.d 54.a 55.c 56.d 57.c 58.b 59.c 60.b 61.c 62.b 63.a 64.c 65.a 66.b 67.b

Chapter 8. Microbial Taxonomy and Diversity

1.b 2.c 3.c 4.a 5.a 6.b 7.c 8.d 9.b 10.a 11.b 12.c 13.d 14.a 15.b 16.a 17.a 18.c 19.c 20.d 21.b 22.a 23.b 24.c 25.a 26.d 27.b 28.a 29.a 30.b 31.c 32.d 33.a 34.a 35.c 36.c 37.a 38.b 39.c 40.a 41.b 42.c 43.d 44.a 45.b 46.c 47.d 48.a 49.b 50.d 51.b 52.a 53.a 54.a 55.d 56.a 57.b 58.d 59.c 60.a 61.b 62.b 63.b 64.a 65.c

Chapter 9. Microbial Structure and Function

1.b 2.c 3.a 4.a 5.a 6.b 7.c 8.a 9.a 10.c 11.a 12.c 13.c 14.d 15.a 16.b 17.a 18.a 19.a 20.b 21.a 22.b 23.b 24.a 25.b 26.a 27.c 28.a 29.b 30.b 31.c 32. D 33.a 34.c 35.d 36.c 37.a 38.c 39.b 40.a 41.a 42.d 43.b 44.c 45.d 46.b 47.a 48.a 49.c 50.c 51.d 52.a 53.c 54.c 55.c 56.a 57.c 58.a 59.c 60.a 61.d 62.d 63.b 64.c 65.d 66.c 67.b 68.c 69.b 70.c 71.a 72.d 73.b 74.d 75.a 76.c 77.b 78.c 79.b 80.d 81.a 82.b

Chapter 10. Microbial Growth and Cultures

1.a 2.c 3.a 4.a 5.b 6.c 7.c 8.b 9.a 10.c 11.d 12.a 13.c 14.d 15.a 16.a 17.a 18.a 19.c 20.d 21.c 22.a 23.a 24.c 25.c 26.d 27.a 28.c 29.a 30.a 31.d 32.c 33.a 34.b 35.c 36.c 37.b 38.a 39.b 40.c 41.d 42.a 43.b 44.d 45.c 46.c 47.b 48.d 49.a 50.c 51.d 52.b 53.b 54.a 55.b 56.a 57.d 58.c 59.b 60.a 61.c 62.b 63.d 64.a 65.c 66.c 67.b 68.c 69.a 70.d 71.c 72.a 73.b 74.b 75.d 76.a 77.c 78.c 79.c 80.d 81.b 82.a 83.b 84.d 85.d 86.c 87.c 88.a 89.a 90.b 91.c 92.c 93.c 94.a 95.c

Chapter 11. Microbial Metabolism

1.a 2.b 3.d 4.b 5.b 6.d 7.a 8.b 9.c 10.a 11.c 12.d 13.c 14.b 15.a 16.c 17.c 18.c 19.d 20.b 21.d 22.a 23.d 24.a 25.b 26.a 27.b 28.a 29.d 30.b 31.c 32.a 33.c 34.a 35.a 36.b 37.a 38.c 39.b 40.c 41.c 42.a

Chapter 12. Microbial Genetics

1.a 2.a 3.b 4.c 5.a 6.b 7.a 8.c 9.a 10.c 11.b 12.d 13.a 14.a 15.b 16.d 17.b 18.a 19.a 20.c 21.a 22.c 23.c 24.b 25.a 26.d 27.c 28.a 29.d 30.a 31.d 32.a

Chapter 13. Microbial Control

1.a 2.b 3.b 4.c 5.a 6.c 7.c 8.a 9.c 10.a 11.c 12.d 13.a 14.b 15.a 16.c 17.c 18.b 19.d 20.c 21.a 22.a 23.a 24.a 25.a 26.a 27.b 28.c 29.b 30.a 31.d 32.c 33.b 34.c 35.d 36.b 37.c 38.b 39.c 40.b 41.a 42.b 43.d 44.b 45.a 46.d 47.a 48.a 49.a 50.b 51.c 52.b 53.d 54.a 55.c 56.d 57.a 58.b 59.a 60.d 61.a 62.b 63.b 64.c

Chapter 14. Microbial Diseases

1.d 2.d 3.a 4.a 5.b 6.d 7.d 8.a 9.c 10.b 11.a 12.c 13.d 14.c 15.a 16.d 17.d 18.a 19.d 20.b 21.b 22.a 23.d 24.c 25.b 26.d 27.b 28.a 29.c 30.a 31.d 32.c 33.b 34.a 35.c 36.d 37.b 38.a 39.d 40.a 41.d 42.b 43.b 44.a 45.c 46.a 47.d 48.a 49.a 50.d 51.b 52.d 53.a 54.c 55.b 56.b 57.a 58.d 59.b 60.c 61.a 62.a 63.a 64.b 65.a 66.c 67.b 68.d 69.a 70.c 71.a 72.c 73.d 74.d 75.c 76.b 77.d 78.b 79.b 80.d 81.a 82.c 83.a 84.b 85.d 86.d 87.b 88.a 89.a 90.b 91.c 92.b 93.d 94.a 95.c 96.c 97.a 98.a 99.b 100.a

www.ingramcontent.com/pod-product-compliance
Lightning Source LLC
Chambersburg PA
CBHW041712210326

41598CB00007B/623